RENEWALS: 691-4574

DATE DUE

MAR 3			
NOV 1 1			
NOV			

WITHDRAWN

UTSA LIBRARIES

D1442815

WITHDRAWN
UTSA LIBRARIES

DESIGN FOR SUCCESS

WILEY SERIES IN SYSTEMS ENGINEERING

Andrew P. Sage

ANDREW P. SAGE AND JAMES D. PALMER
Software Systems Engineering

WILLIAM B. ROUSE
Design for Success: A Human-Centered Approach to Designing Successful Products and Systems

DESIGN FOR SUCCESS

A Human-Centered Approach To Designing Successful Products And Systems

WILLIAM B. ROUSE

A Wiley-Interscience Publication
John Wiley & Sons, Inc.
New York / Chichester / Brisbane / Toronto / Singapore

In recognition of the importance of preserving what has been
written, it is a policy of John Wiley & Sons, Inc., to have books
of enduring value published in the United States printed on
acid-free paper, and we exert our best efforts to that end.

Copyright © 1991 by John Wiley & Sons, Inc.

All rights reserved. Published simultaneously in Canada.

Reproduction or translation of any part of this work
beyond that permitted by Section 107 or 108 of the
1976 United States Copyright Act without the permission
of the copyright owner is unlawful. Requests for
permission or further information should be addressed to
the Permissions Department, John Wiley & Sons, Inc.

Library of Congress Cataloging in Publication Data:
Rouse, William B.
    Design for success: A human-centered approach to successful
    products and systems
      p. cm. -- (Wiley series in systems engineering)
    "A Wiley-Interscience publication."
    Includes bibliographical references and index.
    1. Systems engineering. 2. Human engineering. 3. Engineering
design, I. Title. II. Series.
TA168.R68   1991
620.8'2--dc20
ISBN 0-471-52483-2   90–45055
                                                    CIP

Printed in the United States of America

10 9 8 7 6 5 4 3 2 1

Library
University of Texas
at San Antonio

Contents

Preface

Engineering in the United States is facing an important challenge. While there is no lack of new ideas and inventions, we are having increasing difficulty innovating in the marketplace. Our competitors too often beat us in bringing high-quality new products to market.

It is clear that we need to be more market-driven and user-oriented. We need to understand users' needs and preferences in order to design products and systems that are user-friendly in the broadest sense of the phrase. Few engineering managers and designers would disagree with these observations.

However, in the same breath, they would ask for more than nice words. How does one design products and systems that are market-driven and user-oriented? What *specifically* should one do?

This book answers these questions. The answers are provided in terms of a comprehensive methodological framework for human-centered design of complex systems. This new approach to system design includes four phases: naturalist, marketing, engineering, and sales and service. These four phases cover the entire product life cycle, including

- Understanding users' needs and preferences,
- Conception and market evaluation of alternative ways to satisfy these demands,

- Detailed design and engineering evaluation of product and systems, and
- Fielding and ongoing in-use evaluation.

A wide variety of methods and tools are discussed in the context of this methodological framework. Use of these techniques is illustrated by several case histories that are introduced early in the book and evolve throughout the chapters. The goal is to make human-centered design very concrete and readily applicable to practical and realistically complex design problems. The comprehensive and consistent framework for design, in combination with much "how-to" material and case histories, provides the reader with a self-contained "toolbox" with which to pursue design.

The material in this book is drawn from 10 years of development, utilization, refinement, and extension of concepts and methods. These efforts have occured in commercial and military aviation, the process and power industries, manufacturing, the marine industry, and communications. This wide range of experiences has provided important tests of the usefulness of the concepts in this book. These experiences have also resulted in important changes in the overall approach.

The process of formalizing the approach to design advocated in this book began in the early 1980s. There have been numerous modifications and extensions since then. As a result, the material in this book might not be recognizable to readers of the earliest documents. Nevertheless, the end product represented by this book is the result of an evolutionary process of design, development, and use.

The material in this book has been used in its present form as a basis for a graduate-level course in academia, government, and industry. The experiences of presenting and discussing these concepts, approaches, and methods with students and practitioners have resulted in many important refinements. The purpose of these refinements is to make this book, in itself, more human-centered.

The experience upon which this book is based could not have happened without the sponsorship, partnership, and friendship of many people. Present and former colleagues at Search Technology were central to this process. A wide range of customers and users were also central, not just as sponsors and recipients of results, but also as full partners in the process of understanding and applying human-centered design concepts and methods. The close and committed participation of this range of people is a key element in designing for success.

The evolutionary design of this book has benefited from contributions by a variety of people. Russ Hunt, my partner in the 10-year process of growing and maturing Search Technology, has been a contributor, sounding board, and friendly critic in the development and refinement of the concepts and methods in this book. Bill Cody, Paul Frey, and Ken Boff have been central participants in many discussions about the nature of design and how designers can best be supported. The efficiency with which this book was designed, integrated, and refined is due largely to the efforts of Tena Dumestre. The important role of the aforementioned courses and seminars in testing and shaping the material in this book was enhanced by Sueann Gustin's continuing efforts in coordinating seminar offerings. Finally, I am indebted to the many colleagues who reviewed drafts of the book manuscript and provided very helpful comments and suggestions.

Atlanta, Georgia   William B. Rouse
January 1991

Chapter 1

Introduction

Information technology has enabled the rapid evolution of an increasingly complex society. Worldwide integration of information systems has become an accepted reality. Scheduling and controlling the flows within transportation networks, particularly aviation, have become a necessity. Factory automation has become a key element of competitiveness. And, of course, military systems have become increasingly sophisticated and expensive.

These systems are complex because they have many components, many types of interaction among components, dynamic responses that are difficult to predict, and a wide variety of people interacting to operate, maintain, and manage them. This can lead to considerable uncertainty about the current state of the system, its alternative future states, and how to control the system to achieve desired future states. These uncertainties can make the jobs of people in such systems very difficult, particularly if performance and productivity goals are paramount.

Successful design of systems such as illustrated above presents many challenges. The desired level of integration of these systems requires that subsystems, assemblies, and components be highly compatible, both technically and organizationally. Further, new technologies for system intelligence, for example, require both syntactic and semantic integration—syntactic in the sense that things have to fit together, voltages match, and so forth, and semantic in that computers have to understand the meaning of each other's computations. Finally, integration across the life-cycle of the product or system is needed such that, for example, product design and manufacturing process design can, to the greatest extent possible, occur concurrently.

The scale of these large systems can make project management very difficult. Scheduling, budgeting, and performance monitoring can impose immense "overhead" burdens on system development efforts. As a result, system development can be very expensive and dominated by integration and logistics issues rather than elaboration and proof of the concept being developed.

Finally, the range of use and users of these systems can be quite broad. To accommodate this range, consideration of people's functions in systems has to be carefully factored into the design process. This includes assessing people's abilities, limitations, and preferences relative to their envisioned roles, jobs, and tasks.

HUMAN-CENTERED DESIGN

The trends outlined thus far could lead one to feel that people will increasingly become like "cogs in machines," subjugated to the economic and technological necessities of squeezing the last ounce of performance, and hence competitiveness, out of our systems. There is an element of truth in this perspective, but this vision is no longer a necessity—an alternative design philosophy is possible.

The design philosophy presented in this book has three important attributes. First, this philosophy focuses on the roles of humans in complex systems. Second, design objectives are elaborated in terms of roles of humans. Third, this philosophy specifies design issues that follow from these objectives.

Notice that none of the three elements of this design philosophy focuses on performance, productivity, economics, or related metrics. While these concerns are very important, they subserve human-centered design issues because achieving, for example, performance goals is the consequence of good design rather than the means for creating the design. This distinction becomes clearer if the roles of humans in complex systems are considered.

Roles of Humans

The roles of humans can be addressed in several ways (Rouse, Geddes, and Curry, 1987; Rouse, 1988). There are obvious roles such as operator, maintainer, manager, and designer which denote jobs associated with developing and operating complex systems. Within these roles, people are called upon to exhibit various levels of skill, judgment, and creativity to achieve a range of objectives.

Much more important, however, than abilities to perform, decide, and create is the fact that people are responsible for system objectives at some operational level. It is particularly important that people perceive these responsibilities, willingly accept them, and invest themselves in fulfilling them.

It is imaginable, and in some cases inevitable, that automation of skill, judgment, and perhaps even some aspects of creativity will increasingly be feasible. However, it is not imaginable that automation will ever be legally, ethically, and socially responsible for its actions. Thus, regardless of levels of automation, people will remain responsible for operations of complex systems. People may be geographically remote from these operations and they may be supported by sophisticated computer-based decision aids— nevertheless, they will have ultimate responsibility.

Since people will inevitably be responsible for system operations, it is essential that they perceive the nature of these responsibilities and have appropriate levels of authority to fulfill them. Put simply, people have to be "in charge." Consequently, design philosophies and design methods associated with developing complex systems should explicitly reflect the primacy of supporting people to successfully be "in charge." This conclusion leads to the straightforward assertion that design objectives should be to *support humans to achieve the operational objectives for which they are responsible* (Rouse, 1986, 1988).

This simple assertion has profound implications. From this perspective, the purpose of a pilot is *not* to fly the airplane that takes people from A to B—instead, the purpose of the airplane is to support the pilot who is responsible for taking people from A to B. Similarly, the purpose of factory workers is *not* to staff the factory designed to achieve particular productivity goals—rather, the purpose of the factory is to support the workers who are responsible for achieving productivity goals.

From this viewpoint, the purpose of design is *not* to marshal technology to achieve operational objectives. Instead, design should be oriented toward integrating technology and other resources to support people in ways that are appropriate and conducive to their fulfilling the responsibilities associated with their roles. This is the essence of human-centered design.

Design Objectives

There are three primary objectives within human-centered design. These objectives should drive much of designers' thinking, especially in the earlier stages of design. Case studies and examples in later chapters illustrate the substantial impact of focusing on these three objectives.

The first objective of human-centered design is that it should *enhance human abilities*. This dictates that humans' abilities in the roles of interest be identified, understood, and cultivated. For example, people tend to have excellent pattern recognition abilities. Design should take advantage of these abilities—for instance, by using displays of information that enable users to respond on a pattern recognition basis rather than requiring more analytical evaluation of the information. This example is further discussed in Chapter 8.

The second objective is that human-centered design should help *overcome human limitations*. This requires that limitations be identified and appropriate compensatory mechanisms devised. A good illustration of a human limitation is the proclivity to make errors. Humans are fairly flexible information processors—which is important—but this flexibility can lead to "innovations" that are erroneous in the sense that undesirable consequences are likely to occur.

One way of dealing with this problem is to eliminate innovations, perhaps via interlocks and rigid procedures. However, this is akin to throwing out the baby with the bathwater. Instead, mechanisms are needed to compensate for undesirable consequences without precluding innovations. Such mechanisms, which are discussed in Chapters 7 and 8, represent a human-centered approach to overcoming the human limitation of occasional erroneous performance.

The third objective of human-centered design is that it should *foster user acceptance*. This dictates that users' preferences and concerns be explicitly considered in the design process. Thus, it is important to assure that design results in roles that are meaningful to users. In addition, there are other "stakeholders" in the process of designing, developing, and operating a system—for example, the purchaser or customer who may not be a user. The interests of these stakeholders also have to be considered. Chapters 4 and 10 discuss methods for dealing with issues such as these.

Design Issues

In subsequent chapters, this book presents an overall framework and systematic methodology for pursuing the above three objectives of human-centered design. There are four design issues of particular concern within this framework. The first concern is *formulating the right problem*—making sure that system objectives and requirements are right. All too often, these issues are dealt with much too quickly. There is a natural tendency to "get on with it," which can have enormous negative consequences when requirements are later found to be inadequate or inappropriate.

The second issue is *designing an appropriate solution.* All well-engineered solutions are *not* necessarily suitable. When the three objectives of human-centered design are considered, as well as the broader context within which systems typically operate, it is apparent that the excellence of the technical attributes of a design is necessary but not sufficient to assure that the system design is appropriate and successful.

Given the right problem and appropriate solution, the next concern is *developing it to perform well.* Performance attributes should include operability, maintainability, and supportability—or, using it, fixing it, and supplying it. Supportability includes spare parts, fuel, and most importantly trained personnel—training is discussed in Chapter 9.

The fourth design concern is *assuring user satisfaction.* Success depends on people using the system and achieving the benefits for which it was designed. However, before a system can be used, it must be purchased, which in turn depends on it being technically approved. This process is discussed in Chapter 4.

Figure 1.1 humorously illustrates several of the pitfalls encountered in pursuing these design issues. This cartoon shows many of the obvious things that can go wrong. Clearly, marketing, engineering, manufacturing, and so on, were not communicating in this depiction. One possible cause

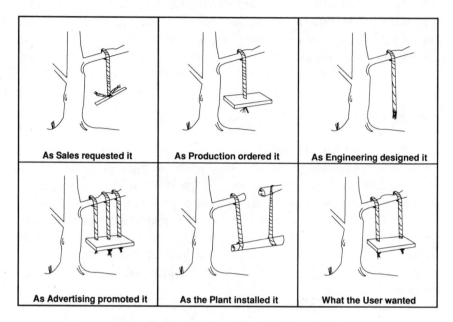

Figure 1.1. Miscommunication in design.

of this miscommunication is the lack of a common framework and perspective for considering and resolving marketing, design, and manufacturing issues. Such a framework is presented in Chapter 2.

Beyond miscommunication or lack of communication, another very important phenomenon that mitigates against human-centered design is displayed in Figure 1.2. The technology spiral is initiated by the confluence of requirements and opportunities. This can result in massive infusions of technology, often with visions of technology as a panacea. This has occurred in the automobile industry in recent years.

Unfortunately, such huge changes can result in the humans in the system being overwhelmed by inadequately planned and poorly explained changes. More fundamentally, operators may have more modes of operation than they know what to do with—this has been the case with many flight management systems for aircraft. Increased functionality for the operator usually results in much more sophisticated maintenance tasks; this is often the case in military systems. Finally, the technology infusions often result in the availability of masses of data about operations, main-

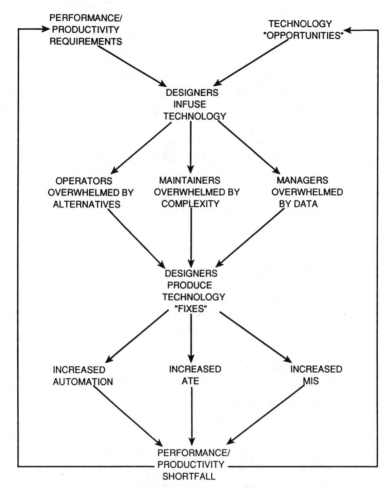

Figure 1.2. The technology spiral.

tenance, performance, and productivity that managers of these systems cannot absorb—information overload seems prevalent in many management domains.

A common reaction to the above problems is to infuse even more technology, as shown in Figure 1.2. This strategy is akin to talking louder when someone who does not speak your language does not understand you. However, the problems are not due to the recipients of the technology.

The problems are due to the ways in which technology opportunities drive the design process.

The result is a performance/productivity shortfall, particularly relative to early inflated expectations. Ironically, this shortfall creates new opportunities and, typically, new technologies have become available offering a new panacea. Consequently, the spiral continues, but performance/productivity goals are still not achieved and, hence, competitiveness problems still persist.

Human-centered design avoids this spiral by focusing on the people throughout a system, the nature of their roles, and how these roles can be supported. Human-centered objectives rather than technology drive the design process. Thus, operators, maintainers, and managers are not overwhelmed because a primary objective of the design process is to avoid this happening.

Levels of Human-Centered Design

Human-centered design is not a new topic. It has been approached on several levels. One level focuses on system users and their activities. Good examples at this level include Card, Moran, and Newell's (1983) studies and models of keystroke selection in word processing tasks. Norman and Draper's (1986) compilation of appropriate metaphors for human–computer interaction and methods of display manipulation is also at this level. This is termed the *interface* level.

Another level emphasizes users' jobs and tasks as they interact with a vehicle, equipment system, or manufacturing process. An example at this *job/task* level is Rasmussen's (1986, 1988) conceptual models of electronic troubleshooting and nuclear power plant control. Another example is Rouse's (1986, 1988) task-oriented methodology for designing aiding systems—this methodology is discussed in Chapter 8.

A third level of human-centered design focuses on the roles of organizations in the life cycle of a system (Booher, 1990). This *organization* level includes many more humans than the "users" of systems. This level is especially important because it defines the context within which the other levels occur.

This book is primarily concerned with the job/task and organization levels of human-centered design. The organization level is considered in terms of both the organization of the design process and the organizational context within which the system being designed will operate. Design and support of people's jobs and tasks are central themes throughout most of this book, including the topics of performance evaluation, job aiding, and training systems. The interface level of human-centered design is not ignored in this volume—however, the primary concern at this level is with determining functional requirements for the interface rather than the specific forms of the displays, input devices, and so on.

DESIGN METHODOLOGY

Concepts such as user-centered design, user-friendly systems, ergonomically designed systems, and so forth have been around for quite some time. Virtually everybody endorses these ideas, but very few people know what to do in order to realize the potential of these concepts. What is needed, and what this book presents, is a methodological framework within which human-centered design objectives can be systematically and naturally pursued.

This framework includes a variety of design procedures that show, in step-by-step fashion, how the design issues discussed earlier can be addressed and resolved. These methods are illustrated in the context of several case studies that are continued throughout the book. These case studies are drawn from actual experiences in aviation, nuclear power, manufacturing, communications, and other domains.

It is very important to emphasize that the methodological framework and design procedures presented here are not prescriptive in the sense that they must be followed in lockstep. Instead, they provide a nominal sequence of steps and guidance that system designers can rely on more or less, depending on their experience and expertise with the issues at hand. Therefore, the comprehensive framework presented in this book is intended to support and augment engineering judgment rather than replace it.

OVERVIEW

The next five chapters focus on elaborating and illustrating (with case studies) a framework for human-centered design and associated methodologies. The essence of this methodology is a set of seven central issues that are addressed in four phases. The motivation for each issue is discussed in great detail, including how it can be addressed and resolved, as well as where in the overall design process it should be addressed, in terms of both planning and execution.

The last four chapters of the book deal with what can be called high-leverage points—aspects of system design where especially high "value added" is possible. The first of these points, system evaluation, is discussed in Chapter 7. The selection of methods and measures is discussed and several case studies are reviewed.

Chapter 8 considers aiding in the sense of providing functionality in the system that directly augments task performance. Aiding is a leverage point when human performance with a baseline system design is not adequate but the basic design is judged to be sound. The question then becomes one of what should be added to assure acceptable performance.

Chapter 9 discusses training as a leverage point. Training is defined as managing peoples' experiences to enhance their potential to perform. Computer-based training is the primary focus in this chapter. Also considered is the fundamental trade-off between training and aiding. Put simply, should one invest in improving people's abilities to perform or should one design more sophisticated systems that do not require talented people? The obvious answer is that one does both, but what should the mix be?

The final chapter concentrates on technology transition. Issues considered include how to get innovative concepts from research and development to design, and how to get innovative products and systems from design to the marketplace. This final leverage point is, of course, the ultimate test of human-centered design.

REFERENCES

Booher, H. R. (Ed.) (1990). *MANPRINT: An approach to systems integration.* New York: Van Nostrand Reinhold.

Card, S. K., Moran, T. P., and Newell, A. (1983). *The psychology of human-computer interaction.* Hillsdale, NJ: Lawrence Erlbaum Associates, Inc.

Norman, D. A., and Draper, S. W. (Eds.) (1986). *User centered system design: New perspectives on human-computer interaction.* Hillsdale, NJ: Lawrence Erlbaum Associates.

Rasmussen, J. (1986). *Information processing and human-machine interaction: An approach to cognitive engineering.* New York: North-Holland.

Rasmussen, J. (1988). A cognitive engineering approach to the modeling of decision making and its organization in process control, emergency management, CAD/CAM, office systems, and library systems. In W. B. Rouse (Ed.), *Advances in man-machine systems research* (Vol. 4, pp. 165–243). Greenwich, CT: JAI Press.

Rouse, W. B. (1986). Design and evaluation of computer-based decision support systems. In S. J. Andriole (Ed.), *Microcomputer decision support systems* (Chapter 11). Wellesley, MA: QED Information Systems.

Rouse, W. B. (1988). Intelligent decision support for advanced manufacturing systems. *Manufacturing Review, 1,* 236–243.

Rouse, W. B., Geddes, N. D., and Curry, R. E. (1987). An architecture for intelligent interfaces: Outline of an approach to supporting operators of complex systems. *Human-Computer Interaction, 3,* 87–122.

Chapter 2

Design and Measurement

What do successful products and systems have in common? The fact that people buy and use them is certainly a common attribute. However, sales is not a very useful measure to designers. In particular, using *lack* of sales as a way to uncover poor design choices is akin to using airplane crashes as a method of identifying design flaws—this method works, but the feedback provided is a bit late.

The question, therefore, is one of determining what can be measured early that is indicative of subsequent poor sales. In other words, what can be measured early to find out if the product or system is unlikely to fly? If this can be done early, it should be possible to change the characteristics of the product or system so as to avoid the predicted failure.

This chapter focuses on the measurement issues that must be addressed and resolved for the design of a new product or system to be successful. Seven fundamental measurement issues are discussed and a framework for systematically resolving these issues is presented. This framework provides the structure within which Chapters 3 through 6 are organized and presented.

MEASUREMENT ISSUES

Figure 2.1 presents seven measurement issues that underlie successful design (Rouse, 1987). The "natural" ordering of these issues depends on

Viability ———→Are the Benefits of Systems Use
Sufficiently Greater than Its Costs?

Acceptance ——→Do Organizations/Individuals Use the
System?

Validation ———→Does the System Solve the Problem?

Evaluation ———→Does the System Meet Requirements?

Demonstration –→How Do Observers React to System?

Verification ———→Is the System Put Together as Planned?

Testing ———→Does the System Run, Compute, Etc.?

Figure 2.1. Measurement issues.

one's perspective. From a nuts and bolts engineering point of view, one might first worry about testing (i.e., getting the system to work), and save issues such as viability until much later. In contrast, potential users and customers are usually first concerned with viability and only worry about issues such as testing if problems emerge.

A central element of human-centered design is that designers should address the issues in Figure 2.1 in the same order that users and customers address these issues. Thus, the *last* concern is, "Does it run?" while the *first* concern is, "What matters?" or what constitutes benefits and costs?

Viability

Are the benefits of system use sufficiently greater than the costs? While this question cannot be answered empirically prior to having a design, one can determine how the question is likely to be answered.

How do potential users and customers characterize benefits? Are they looking for speed, throughput, an easier job, or appealing surroundings? What influences their perceptions of these characteristics?

How do potential users and customers characterize costs? Is it simply purchase price? Or, do costs include the costs of maintenance and perhaps training? Are all the costs monetary? Are there cognitive costs of learning and psychological costs of having to change ways of doing things? What influences peoples' perceptions of these types of cost?

Chapters 3 and 4 discuss how these questions can be answered. At this point, consider what happens if one does *not* answer them. It is quite possible that one will design a system that runs well, is put together as planned, and passes requirements evaluation with flying colors. However, it may be too expensive, too difficult to learn to use, or perhaps solve a problem that an insufficient number of people perceive to be important. As a result, a first-rate piece of engineering design may not be viable.

Acceptance

Do organizations/individuals use the system? This is also a question that cannot be answered definitively before having the results of design. But one can determine in advance the factors that are likely to influence the answer.

Products and systems face acceptance problems when they are perceived not to function properly, to require too much change of habits, and do not fit into the current way of doing things. Consequently, answers to several questions should be sought early in the design process.

What are potential users' and customers' perceptions of the credibility of the technologies that are likely to underlie the new product or system? Are they concerned about performance and reliability? Or do they have a vague feeling of uncertainty?

Will the new system require any procedural or organizational changes? How are people likely to feel about these changes? Are people likely to be uncertain about their roles, or perhaps even the future of their jobs?

How do the organizations of potential users and customers currently function? Will the new system have to displace a strong or a weak incumbent? Does the organization have any brand loyalties?

Chapters 4 and 10 discuss these and related questions in detail. Consider the risks if one does *not* pursue answers to these acceptance questions. Clearly, lack of acceptance can undermine sales. This may, however, occur in a very subtle way. The product or system may be initially perceived as viable, resulting in several customers purchasing it. Subsequently, users may balk because, for example, their roles are changed in very undesirable ways. These users may make their feelings widely known. As a result, the system may never get a chance to make a second impression.

Validation

Does the system solve the problem? This, or course, leads to the question "What is the problem?" The nature of this question was discussed in Chapter 1.

A primary difficulty with this question is that potential users and customers are often quite willing to say what the problem is. However, they are not necessarily correct. Often it takes a great deal of interaction to uncover the underlying problem that is causing, for example, performance or productivity symptoms. This interaction process is discussed in Chapter 5.

Thus, the customer is not necessarily always right! If the issue is one of preferences, then of course the customer is, by definition, right. But if the issue is one of fact (e.g., causality of problems), then everyone has an equal opportunity to be wrong. If the designer perceives that stated facts are incorrect, then it is his or her responsibility to help correct them.

Once the right problem has been identified, the next question concerns finding valid solutions. This involves asking the question, "How would we know if the problem was solved?" One answer is, "Everybody buys it." However, as emphasized earlier, this answer—sales—comes much too late to be of use in design. Accordingly, intermediate measures are needed such as potential users' and customers' perceptions of the likely validity of the solution. Examples of this approach to measurement are discussed in Chapter 4.

Evaluation

Does the system meet requirements? Formulation of the design problem should result in specification of requirements that must be satisfied for a design solution to be successful. Examples include speed, accuracy, throughput, and manufacturing costs.

Evaluation is concerned with assessing these characteristics of a design. This assessment can occasionally be performed analytically—via pencil and paper or, more likely, via computer programs. For complex systems involving significant levels of human–system interaction, analytical evaluation is seldom sufficient. Empirical assessment becomes necessary.

Of course, the ultimate empirical evaluation is sales of the product or system. But, as emphasized repeatedly, this type of experiment runs the risk of irrecoverable failure. Therefore, empirical evaluation should occur much earlier. Evaluation methods and measures are discussed in Chapter 7.

Demonstration

How do observers react to the system? It is very useful to get the reactions of potential users and customers long before the product or system is ready for evaluation. It is important, however, to pursue demonstration in a way that does not create a negative first impression.

This risk can often be minimized by using "surrogate" users and customers. Such surrogates can often be one or more people from within one's own organization whose experience and expertise make them good models of the targeted users and customers. If such people are not available, it is quite likely that local consultants can be found—for example, retired pilots or power plant operators.

Open-ended demonstrations can be very useful early in the process. However, as ideas mature, it is helpful to provide some structure in terms of questionnaires, rating forms, or checklists that surrogate users and customers are asked to complete. For later demonstrations, the actual target population should be shown the current design and asked to respond.

Verification

Is the system put together as planned? This question can be contrasted with a paraphrase of the validation question: Is the plan any good? Thus, verification is the process of determining that the system was built as intended, but does not include the process of assessing whether or not it is a good design.

In the software industry, there are standard approaches to verification including third-party audits, talk-throughs, and walk-throughs. It tends to be a time-consuming but nonetheless important process. Verification is briefly considered again in Chapter 5.

Testing

Does the system run, compute, and so forth? This is a standard engineering question. It involves issues of physical measurement and instrumentation for hardware, and runtime inspection and debugging tools for software. The role of testing is discussed in Chapter 5, although testing methods and procedures are not discussed in this book.

Overall Approach

The remainder of this chapter presents a framework within which the seven measurement issues can be pursued. A recurring theme in the elaboration of this framework is the way in which Figure 2.1 should be addressed. The overall approach is that measurement should be planned top-down (i.e., starting with viability) and executed bottom-up (i.e., starting with testing). Thus, as noted earlier, one first asks, "What matters?" and last asks, "Does it run?" However, the definitive answers to these questions are obtained in the opposite order.

The concept is quite simple. But it begs the question of exactly what is meant by planning and execution. This requires that the seven measurement issues be considered in the context of system design and system usage.

MEASUREMENT AND SYSTEM DESIGN

Figure 2.2 presents a *very* simplified view of the process of system design as typically conceptualized by many people. It is simplified in the sense that the process is depicted as much too linear and too orderly. In fact, reality involves much iteration and a great amount of parallelism as later discussions illustrate. Nevertheless, at this point the general sequence from needs to solutions provides the necessary structure for illustrating the relationships between the design process and the measurement issues in Figure 2.1.

Figure 2.3 shows how the issues from Figure 2.1 tend to be integrated within the process of Figure 2.2. This illustration depicts measurements in terms of feedback loops from products to needs, problem formulations, requirements, and solution concepts. This type of representation provides a useful means for differentiating the planning and execution of measurements.

The *planning* of measurements should proceed from the outer loops of Figure 2.3. (In other words, as noted earlier, planning should start at the top of the list in Figure 2.1.) Therefore, one should determine very early

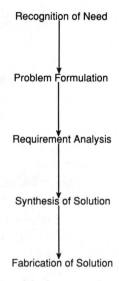

Figure 2.2 Simplified system design process.

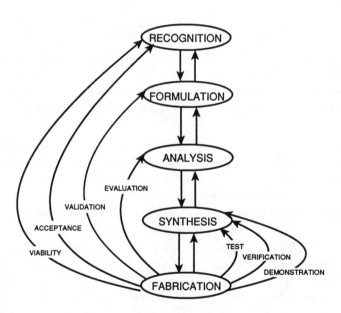

Figure 2.3. Integration of measurement issues and system design.

in the design process how viability, acceptability, and validity will be assessed. More specifically, what methods and measures will be used to make these assessments long before product sales, or equivalent, provide an indication of the success or failure of the product?

The *execution* of measurements should, and often must, proceed from the inner to outer loops of Figure 2.3 or, equivalently, start at the bottom of the list in Figure 2.1. Once a product or prototype emerges from fabrication, one can then execute the plans formulated earlier. In the absence of plans, measurement issues may be pursued in an inappropriate or inefficient manner or, more unfortunately, not pursued at all.

The conclusions drawn from Figure 2.3 provide important elements of the comprehensive approach to measurement that is presented later in this chapter. These conclusions are not particularly elaborate or subtle—one might reach similar conclusions from a careful analysis of planning and executing a vacation or a gourmet meal. However, these conclusions, when combined with a few observations regarding system usage, suggest very substantial changes of traditional ways of viewing system design.

MEASUREMENT AND SYSTEM USAGE

It is important to recognize that *all* of the issues in Figure 2.1 are inevitably addressed. If designers do not address some of the measurement issues, then users eventually will, sometimes with unfortunate results. Of course, the primary concerns of system users do not include measurements per se, but rather involve performing tasks of interest. Nevertheless, system usage is likely eventually to uncover prior inadequate or inappropriate resolution of the seven measurement issues.

System usage concerns are summarized in Figure 2.4. The wording in this figure was chosen to illustrate usage of information systems. Different task contexts might require rewording the questions, but the nature of the questions would remain the same.

A comparison of Figures 2.1 and 2.4 provides an interesting contrast in "tones." In Figure 2.1, the designer is asking whether or not the characteristics of the design solution are appropriate, adequate, and so on. On the other hand, the user is asking (in Figure 2.4) whether or not the characteristics of the design solution provide appropriate and adequate benefits. Put simply, designers tend to be interested in the features of their designs, while users are interested in the benefits of the features.

Needs ⟶ Do I Know What I Want?

Requests ⟶ Can I Communicate My Needs?

Responses ⟶ Does the System Provide Substantive Responses?

Usability ⟶ Can I Utilize These Responses?

Impact ⟶ Does Usage Help Satisfy My Needs?

Value ⟶ Is This Help Worth the Price?

Figure 2.4. System usage concerns.

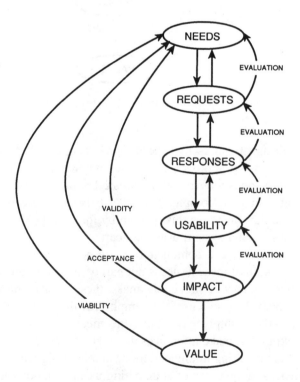

Figure 2.5. Integration of measurement issues and system usage.

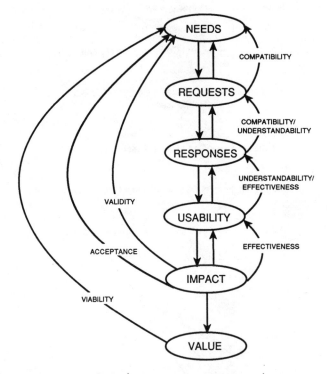

Figure 2.6. Integration of measurement issues and system usage.

Successful design requires that the designer adopt both perspectives. Thus, a comprehensive approach to measurement requires that the designer ask all of the questions in both Figures 2.1 and 2.4. Obviously, there is a significant (but not complete) overlap between these two sets of concerns. Figure 2.5 illustrates this overlap by integrating a subset of the measurement issues and system usage concerns.

Testing, verification, and demonstration are not shown in Figure 2.5. Hopefully, resolution of these issues is not part of system usage, although, as noted earlier, users will inevitably make these measurements if designers avoid them. It is also useful to emphasize that users seldom plan measurements—they simply experience the measurement process and, implicitly or explicitly, make inferences that relate to measurement issues. To assure that the resulting inferences are desirable, designers should plan users' experiences in the sense of anticipating users' reactions and choosing system functionality and features accordingly.

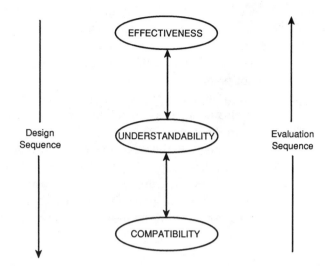

Figure 2.7. Top-down design vs. bottom-up evaluation (Rouse, 1984,1986).

LEVELS OF EVALUATION

Evaluation appears several times in Figure 2.5. It is useful to decompose these instances of evaluation into three component issues (see below), as shown in Figure 2.6. The need for a finer-grained view of evaluation emerged during several system evaluation efforts in the aerospace and nuclear power industries (Rouse, 1984, 1986; Rouse et al., 1984). These experiences are discussed in Chapter 7.

The three evaluation issues in Figure 2.6 are defined as follows. *Compatibility* is the extent to which the nature of physical presentations to the user and the responses expected from the user are compatible with human input–output abilities and limitations. *Understandability* is the extent to which the structure, format, and content of the user–system dialogue result in meaningful communication. *Effectiveness* is the extent to which the system leads to improved performance, makes a difficult task easier, or enables accomplishing a task that could not otherwise be accomplished.

The order in which these issues are considered depends on one's perspective, as illustrated in Figure 2.7. Design begins by considering the requirements for an effective solution, proceeds to being concerned with the understandability of human–system communication, and finally focuses on the compatibility of displays, input devices, and the environment.

Evaluation, on the other hand, inevitably must first deal with compatibility. If a user cannot see the displays or reach the keyboard, then understandability and effectiveness are moot points. Similarly, if a user can see the display but labels, symbols, and so forth, are meaningless, then effectiveness is a moot point.

I have repeatedly seen instances of potentially very effective products and systems failing evaluation due to compatibility and understandability problems. This is due to the natural tendency in design to put the lion's share of the resources, and sometimes all the resources, into assuring effectiveness. However, in these experiences, I found that potential users and customers never got to see the fruits of this investment because they got bogged down, and turned off, by unresolved compatibility and understandability deficiencies. As a result, the designers never had the opportunity to make a second impression.

A FRAMEWORK FOR MEASUREMENT

The discussion thus far has served to emphasize the diversity of measurement issues from the perspectives of both designers and users. If each of these issues were pursued independently, as if they were ends in themselves, the costs of measurement would be untenable. Yet, each issue is important and should not be neglected.

What is needed, therefore, is an overall approach to measurement that balances the allocation of resources among the issues of concern at each stage of design. Such an approach should also integrate intermediate measurement results in a way that provides maximal benefit to the evolution of the design product. These goals can be accomplished by viewing measurement as a process involving the four phases shown in Figure 2.8.

A measurement plan should be developed that explicitly recognizes these four phases, anticipates the specific methods and measures that will be used to address each issue in each phase, and considers the decision-making criteria that will be used to judge the outcomes of each measurement effort. The relative emphasis on each phase and issue will depend on the nature of the design effort—for example, evolutionary vs. revolutionary product changes. Nevertheless, all four phases and all seven measurement issues should be explicitly considered, if only to document the fact that a particular phase and/or issue has, in effect, already been resolved.

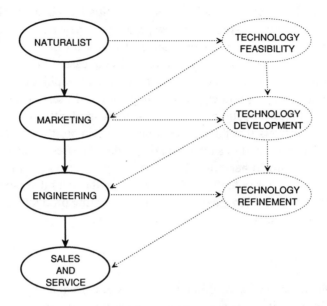

Figure 2.8. A framework for measurement

Naturalist Phase

The term "naturalist" was chosen for the initial phase to emphasize particular characteristics of this phase. A naturalist is an individual who studies phenomena in their natural surroundings. A naturalist watches and listens, but does not try to influence the phenomena of interest. From a design perspective, naturalism involves understanding the world as it is, rather than attempting to introduce changes to the world—such efforts are initiated during the marketing phase.

The naturalist phase involves understanding the domains and tasks of users from the perspective of individuals, the organization, and the environment. This understanding not only includes users' activities, but also the prevalent values and attitudes relative to productivity, technology, and change in general. Evaluative assessments of interest include identification of difficult and easy aspects of tasks, barriers to and potential avenues of improvement, and the relative leverage of the various stakeholders in the organization.

The need for these assessments is easy to justify when entering a new market—attempting to provide a new product to a new population of users. Only the hopelessly naive would suggest that a bit of naturalism is not required in this situation. The more frequent cases, however, involve attempting to provide a new product to a familiar population of users, or provide an existing product to a new population of users. There is a great tendency in these situations to assume, usually implicitly, that one already knows the answers that the naturalist phase would produce. This tendency may have contributed to such dramatic product failures as the Edsel, New Coke, and many others. It seems reasonable to suggest that the only situation where it is potentially sensible to skip the naturalist phase is when one is trying to develop a new version of an existing product for its current population of users.

Marketing Phase

Once one understands the domain and tasks of current and potential users, one is in a position to conceptualize alternative products or systems to support these users. Product concepts can be used for initial marketing in the sense of determining how users react to the concepts. Users' reactions are needed relative to validity, acceptability, and viability. In other words, one wants to determine whether or not users perceive a product concept as solving an important problem, solving it in an acceptable way, and solving it at a reasonable cost.

More specifically, questions related to validity are concerned with the accuracy of the conclusions drawn from the naturalist phase. Acceptability may, at first glance, seem quite simple, but one quickly realizes that acceptance is multidimensional and subtle. Therefore, a fairly structured approach to measurement is helpful. Viability concerns how much one would be willing to pay for the product, who would make the ultimate purchase decision, and what process is used to recommend and justify product purchases and system acquisitions.

While all of the validity, acceptability, and viability questions cannot be resolved during the marketing phase, one can test and elaborate plans for eventual resolution of these issues. Further, one can test and refine the hypotheses that emerged from the naturalist phase regarding the appropriate impact of user characteristics on design choices and product concepts. Thus, this phase produces both an assessment of the relative merits of the multiple product concepts that have emerged up to this point, as well as a preview of any particular difficulties that are likely when one later tries to assure that users perceive the resulting product as valid, acceptable, and viable.

Engineering Phase

One now is in a position to begin trade-offs between desired conceptual functionality and technological reality. As indicated in Figure 2.8, technology development will usually have been pursued prior to and in parallel with the naturalist and marketing phases. This will have at least partially assured that the product concepts shown to potential users and customers were not technologically or economically ridiculous. However, one now must be very specific about how desired functionality is to be provided, what performance is possible, and the time and dollars necessary to provide it.

Most of the effort in this phase is associated with using various design methodologies to transform conceptual designs to detailed designs. Testing and verification will be pursued in the process of producing an operational prototype. Demonstrations of prototypes are an important way to get early feedback, perhaps initially from surrogate users rather than the targeted users. The evaluation issues of compatibility, understandability, and effectiveness are pursued and resolved in this phase.

Sales and Service Phase

As this phase begins, the product should have successfully been tested, verified, demonstrated, and evaluated. (Of course, if these earlier efforts have been unsuccessful, sales and service may never be an issue!) From a measurement point of view, the focus is now on validity, acceptability, and viability. It is also at this point that one assures that implementation conditions are consistent with the assumptions underlying the design basis of the product or system.

If the earlier naturalist and marketing phases were well done, the sales and service phase should not be calamitous. Nonetheless, the final marketing, sales, installation, and ongoing service of a product or system are the activities that should provide the validation, acceptance, and viability measurements planned earlier. To an extent, sales volume (e.g., number of units bought) can be viewed as a measure of these three attributes. However, the single measure of sales volume is very much inadequate, and usually much too late, to serve as a means for diagnosing and remediating problems.

The sales and service phase of measurement is important even if the product is "presold"—for example, delivered as the result of a contracted R&D effort. If the goal is for the product or system to be accepted and used as intended, then marketing in one form or another has to continue throughout the product life cycle. During installation and service, one can maintain, enhance, and extend the relationships with users developed during earlier phases.

Continued contact with users and customers provides both formal and informal means for assessing validity, acceptability, and viability. This contact also substantially lessens the investments necessary for subsequent naturalist and marketing phases for new products. In particular, recognition and discovery of new problems and needs can serve as important inputs to new product and service ideas.

The Role of Technology

It is important to note the role of technology in the human-centered design process. As depicted in Figure 2.8, technology is pursued in parallel with the four phases of the design process. In fact, technology feasibility, development, and refinement usually consume the lion's share of the resources in a product or system design effort. But, as noted repeatedly, technology should not drive the design process. Human-centered design objectives should drive the process and technology should support these objectives.

ORGANIZATION FOR MEASUREMENT

Figure 2.9 illustrates how the seven measurement issues should be organized, or sequenced, in the four phases. *Framing* an issue denotes the process of determining what an issue means within a particular context and defining the variables to be measured. *Planning* is concerned with devising a sequence of steps and schedule for making measurements. *Refining* involves using initial results to modify the plan, or perhaps even rethink issues and variables. Finally, *executing* is the process of making measurements and interpreting results. This whole process is elaborated in Chapters 3 through 6.

PHASE / ISSUE	NATURALIST	MARKETING	ENGINEERING	SALES AND SERVICE
Viability	Frame	Plan	Refine	Execute
Acceptance	Frame	Plan	Refine	Execute
Validity	Frame	Plan	Refine	Execute
Evaluation	--------	Frame/Plan	Refine/Execute	--------
Demonstration	--------	Frame/Plan	Refine/Execute	--------
Verification	--------	Frame/Plan	Refine/Execute	--------
Testing	--------	Frame/Plan	Refine/Execute	--------

Figure 2.9. Organization of measurement process.

Figure 2.9 provides a useful context in which to discuss typical measurement problems. There are two classes of problems of interest. The first class is planning too late, where, for example, failure to plan for assessing acceptance can preclude measurement prior to putting a product into use. The second class of problems is executing too early, where, for instance, demonstrations are executed before resolving test and verification issues, and potentially lead to negative initial impressions of a product or system.

Reflecting upon the line of reasoning advanced in this chapter, many of the points are quite simple and obvious. However, there are alternative points of view. Predominant among these alternatives is the view that technology issues, from which products and systems will "emerge," must be addressed first.

This point of view is not inherently wrong and, as noted earlier, technology usually should proceed in parallel to the efforts advocated in this chapter. But a technology-driven approach tends to consume many resources and often precludes adequate and appropriate measurement. As shown in Figure 2.10, the results of inadequate/inappropriate measurement range from lost investments to failed ventures as users and customers

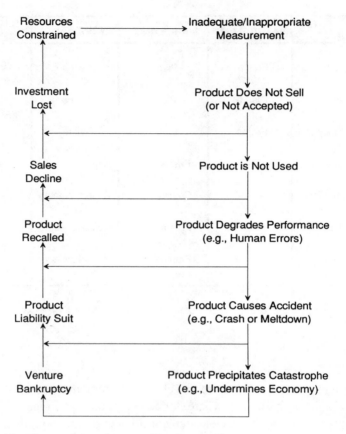

Figure 2.10. The causes and costs of inadequate/inappropriate measurement

inevitably make the measurements that designers neglected to make. Further, as depicted in this figure, a cycle can result whereby inadequate/inappropriate measurement yields poor products and systems that cause decreased sales and depleted resources, which precipitates inadequate/inappropriate measurement, and so on.

Planned measurement, in the manner advocated in this chapter, is the most economical way to proceed. Such planning should enable producing viable, acceptable, and valid products and systems within reasonable schedule and budget constraints, and with relatively few unfortunate surprises.

REFERENCES

Rouse, W. . (1984). Computer-generated display system guidelines. Volume 2: *Developing an evaluation plan* (Rept. NP-3701, Vol. 2). Palo Alto, CA: Electric Power Research Institute.

Rouse, W. B. (1986). Design and evaluation of computer-based decision support systems. In S. J. Andriole (Ed.), *Microcomputer decision support systems* (Chapter 11). Wellesley, MA: QED Information Systems.

Rouse, W. B. (1987). On meaningful menus for measurement: Disentangling evaluative issues in system design. *Information Processing and Management, 23,* 593–604.

Rouse, W. B., Kisner, R. A., Frey, P. R., and Rouse, S. H. (1984). *A method for analytical evaluation of computer-based decision aids* (Rept. NUREG/CR-3655). Oak Ridge, TN: Oak Ridge National Laboratory.

Chapter 3

The Naturalist Phase

The purpose of the naturalist phase is gaining an understanding of users' domains and tasks. This includes assessing the roles of individuals, their organizations, and the environment. Also of interest is identifying barriers to change and avenues for change.

The result of the naturalist phase is a formal description of users, their tasks, and their needs. This description can take many forms, ranging from text to graphics and ranging from straightforward descriptions to theories and hypotheses regarding users' behaviors. This range is illustrated in the case studies introduced in this chapter.

This chapter elaborates and illustrates the process of developing descriptions of users, tasks, and needs. The descriptions resulting from the naturalist phase are the starting point for the marketing phase, which is discussed in Chapter 4.

IDENTIFYING THE USER

Who is the user? This is *the* central question with which a human-centered design effort should be initiated. The answer to this question is *not* sufficient for success—this issue is discussed in Chapters 4 and 10. However, the answer to this question is certainly necessary.

User Populations

The key issue is identifying a set of people whose tasks, abilities, limitations, attitudes, and values are representative of the total user population of interest. Since any single user or group of users will obviously reflect their own organizations, it is usually necessary to sample multiple organizations to understand the overall population of interest.

An exception to this guideline occurs when the total population of users resides in a single organization—for example, Air Force pilots. At an extreme, I can recall a design effort to develop a data entry system where we only queried and studied a single user because that individual was the only potential user of the system.

Designers as User Surrogates

Rather than explicitly identifying users, it is common for designers to think, perhaps only tacitly, that they understand users and, therefore, can act as users' surrogates. To the extent that designers are former users, this approach has some value. However, it is inherently limited from capturing the abilities, attitudes, and aspirations of current or potential users, as well as the current or potential impact of their organizations.

For example, the case studies that are discussed later in this chapter involve domains and technologies with which my colleagues and I have considerable experience and formal training. It probably would have been quite easy to defend an assertion that we understood the users, tasks, and so forth, and did not need to devote attention to the naturalist phase. But, as is later discussed, we did pursue the naturalist phase for these efforts and learned how little we really knew beforehand. This knowledge, and a few midcourse corrections enabled by the framework introduced in Chapter 2, prevented some potentially unfortunate results.

Elusive Users

It is often argued, particularly for advanced technology efforts, that the eventual users of the product or system of interest do not yet exist—there are no incumbent users. This is very seldom true because there are actually extremely few products and systems that are designed "from scratch." Even, for example, when designing the initial spacecraft, much was drawn from previous experiences in aircraft and submarines.

For those very rare cases where a truly new system is being designed, one still should identify the population from which users will be drawn. Thus, it should be possible to characterize likely users' abilities, limitations, and preferences. For example, regardless of what systems the military envisions, they know that the population from which they can recruit users over the next 18 years is already born. The population available over the next five years is defined even more crisply in terms of the extent and nature of their education, attitudes toward work and the military, and likely aspirations.

Therefore, the argument that one is inherently limited from identifying and understanding the likely users of products and systems is a fairly weak argument. Of course, characterizing hypothetical users can present problems in the sense that the appropriate impact of these characteristics on design choices may be far from clear.

For instance, consider the likely users of a future product or system, for example, a data-base system for consumers. It is easy to imagine these users being characterized as watching much television, not liking to read, playing many computer games, and having poor math skills. While this characterization is of some direct use, it primarily serves to generate hypotheses—for example, emphasize spatial rather than verbal images and assume reasonable skills with keyboards and joysticks. These hypotheses can be evaluated during the marketing phase.

METHODS AND TOOLS FOR MEASUREMENT

How does one identify users and, in particular, how does one determine their needs, preferences, values, and so on? Observation is, of course, the necessary means. Initially, unstructured direct observation may be appropriate. Eventually, however, more formal means should be employed to assure unbiased, convincing results. Figure 3.1 lists the methods and tools appropriate for answering these types of questions.

Magazines and Newspapers

To gain an initial perspective on what is important to a particular class of users or a particular industry, one should read what they read. Trade magazines and industry newspapers publish what interests their readers. One can capitalize on publishers' insights and knowledge by studying articles for issues and concerns. For example, is cost or performance more important? Is risk assessment, or equivalent, mentioned frequently?

One should pay particular attention to advertisements, because advertisers invest heavily in trying to understand customers' needs, worries, and preferences. One can capitalize on advertisers' investments by studying the underlying messages and appeal in advertisements.

It is useful to create a file of articles, advertisements, brochures, catalogs, and other such matter that appear to characterize the users of interest. The contents of this file can be slowly accumulated over a period of many months before it is needed. This accumulation might be initiated in light of long-term plans to move in new directions. When these long-term plans become short-term plans, this file can be accessed, the various items juxtaposed, and an initial impression formed relatively quickly.

Data Bases

There are many relatively inexpensive sources of information about users available via online data bases. With these sources, a wide variety of

METHODS AND TOOLS	PURPOSE	ADVANTAGES	DISADVANTAGES
Magazines and Newspapers	Determine customers' and users' interests via articles and advertisements.	Use is very easy and inexpensive.	Basis and representativeness of information may not be clear.
Data bases	Access demographic, product, and sales information.	Coverage is both broad and quantitative.	Available data may only roughly match information needs.
Questionnaires	Query large number of people regarding habits, needs, and preferences.	Large population can be inexpensively queried.	Low return rates and shallow nature of responses.
Interviews	In-depth querying of small number of people regarding activities, organization, and environment.	Face-to-face contact allows in-depth and candid interchange.	Difficulty of gaining access, as well as time required to schedule and conduct.
Experts	Access domain, technology, and/or methodological expertise.	Quickly able to come up to speed on topics.	Cost of use and possible inhibition on creating in-house expertise.

Figure 3.1. Methods and tools for measurement.

questions can be answered. How large is the population of users? How are they distributed, organizationally and geographically? What is the size of their incomes? How do they spend it?

Such data bases are also likely to have information on the companies whose advertisements were identified in magazines and newspapers. What are their sales and profits? What are their patterns of growth?

By pursuing these questions, one may be able to find characteristics of the advertisements of interest that discriminate good vs. poor sales growth and profits. Such characteristics might include leading-edge technology, low cost, and/or good service.

Questionnaires

Once magazines, newspapers, and data bases are exhausted as sources of information, attention should shift to seeking more specific and targeted information. An inexpensive approach is to mail, or otherwise distribute, questionnaires to potential users to assess how they spend their time, what they perceive as their needs and preferences, and what factors influence their decisions.

Questions should be brief, have easily understandable responses, and be straightforward to answer. Multiple-choice questions or answers in terms of rating scales are much easier to answer than open-ended, essaylike questions, even though the latter may provide more information.

Low return rate can be a problem with questionnaires. Incentives can help. For example, those who respond can be promised a complete set of the results. We got an excellent response rate for one effort where a few randomly selected respondents were given tickets to Disney World.

Results with questionnaires can sometimes be frustrating. Not infrequently, analysis of the results leads to new questions that one wishes had been on the original questionnaire. These new questions can, however, provide the basis of a follow-up agenda.

Interviews

Talking with users directly is a rich source of information. This can be accomplished via telephone, but face-to-face is much better. We have found the use of two interviewers to be invaluable to enable one person to maintain eye contact and the other to take notes. The use of two interviewers also later provides two interpretations of what was said.

Usually, interviewees thoroughly enjoy talking about their jobs and what types of products and systems would be useful. We have been repeatedly surprised by the degree of candor people exhibit. Because of this, interviewees usually do not like their comments tape-recorded.

It is helpful if interviewees have first filled out questionnaires, which can provide structure to the interview as they explain and discuss their answers. Questionnaires also assure that they will have thought about the issues of concern before the interview. In the absence of a prior questionnaire, the interview should be carefully structured to avoid unproductive tangents. This structure should be explained to interviewees before beginning the interview.

Experts

People with specialized expertise in the domain of interest, the technology, and/or the market niche can be quite valuable. Persons who were formerly users within the population of interest tend to be especially useful. These people can be accessed via informal telephone calls (which are surprisingly successful), gathered together in invited workshops, and/or hired as consultants.

While experts' knowledge can be essential, it is very important that the limitations of experts be realized. Contrary to the demeanor of many experts, very few experts know everything! Listen and filter carefully.

Furthermore, it is very unlikely that one expert can cover a wide range of needs. Consider multiple experts. This is not owing to a need to get a good average opinion. It is due to the necessity to cover multiple domains of knowledge.

Summary

The success of all of the above methods and tools depends on one particular ability of designers—the ability to *listen*. During the naturalist phase, the goal is understanding users rather than convincing them of the merits of particular ideas or the cleverness of the designers. Designers will get plenty of time to talk and expound in later phases of the design process. At this point, however, success depends on listening.

CASE STUDIES

Quite a few case studies are discussed in this book. They serve to make the somewhat abstract human-centered design framework much more concrete. Three case studies are introduced in this chapter and are subsequently revisited in Chapters 4, 8, and 9. Other, briefer descriptions of case studies are presented and discussed in Chapters 5 through 7.

The three case studies discussed in this chapter are tersely described in Figure 3.2. The domains of these studies range from aircraft crew systems in particular, to aerospace crew systems in general, to technically oriented jobs in general. The range of these domains, as well as the many other domains discussed in other chapters, illustrates the wide applicability of the human-centered design concepts presented in this book.

Intelligent Cockpit

The problem that motivated this design effort was the emerging information overload experienced by aircraft pilots as advanced information technology was (and is) used to make it possible for aircraft to meet increased performance and safety goals. The infusion of this information technology threatens to create the proverbial situation where the cure is worse than the disease.

CASE STUDY	DOMAIN	ISSUES
Intelligent Cockpit	Aircraft crew systems.	Decision aiding and other support to improve performance and safety.
Design Information System	Crew systems for aerospace vehicles and equipment systems.	Support for accessing and utilizing technical information during design.
Trade-off Analysis Tool	Operations, maintenance, and other jobs in a wide range of equipment and organizational systems.	Identification and evaluation of alternative approaches to training and aiding.

Figure 3.2. Three case studies.

The need expressed by the customer was to develop an integrated, human-centered cockpit concept that would enable pilots to thrive in the likely information-rich aviation environments of the future. In addition, there was a strong desire for the new cockpit concept to provide a convincing demonstration of the value of state-of-the-art intelligent systems technology, a subset of which is termed artificial intelligence technology.

Several issues were prominent in this effort. First, it was very clear who the user population was. They tend to wear uniforms with wings over their left shirt pockets.

However, it was less clear that this population should dictate the design solution. While pilots are very knowledgeable of "what is," they are less knowledgeable and fairly conservative about "what might be."

Another concern was the extent to which current aircraft systems should shape and constrain our thinking. The evolutionary vs. revolutionary trade-off was a topic of much debate early in this project.

Technological risk was clearly a driving issue in this effort. Could an intelligent cockpit be made to work and meet performance expectations? While cost was a concern, it was much less central than this question.

The payoff sought in developing an intelligent cockpit was improved performance and safety rather than cost savings. However, it was easy to imagine an intelligent cockpit enabling reduced crew size which would result in substantial savings. Nevertheless, this possibility was given scant attention during this project.

Design Information System

The problem motivating this design effort was the great difficulty designers experience in accessing and utilizing the wide variety of technical information that is relevant and important in design. The current levels of difficulty in accessing as well as utilizing technical information, particularly across design disciplines, result in less than fully informed design decision making.

A primary issue in this effort was identifying potential users for design information systems. Many people intentionally influence the function and form of design products and systems, and design decision making is pervasive (Rouse, 1987a). Consequently, central questions are who to support, where, and when?

Another issue of great concern was the feasibility of compiling and providing on-line the extensive data bases and knowledge bases necessary to achieving significant improvements in the access and utilization of technical information. Many government and industrial efforts are accelerating the process of creating computer-accessible databases and knowledge bases. But the scope of this overall effort is enormous and, relative to this scope, progress is fairly slow.

The payoff sought by this effort to develop an information system was more effective and efficient design decision making. Such an impact would, hopefully, yield better-performing, higher-quality products and systems at lower costs.

Trade-off Analysis Tool

This design effort was motivated by the potential to improve the way in which training and aiding systems are designed, and training vs. aiding trade-offs are resolved. These trade-offs address the question of whether people should receive enhanced training to be able to deal with increasingly sophisticated technology, or whether the technology should be made sufficiently "smart" to enable the use of personnel with lower levels of knowledge and skills.

The answer is probably some mix of training and aiding, the characteristics of which vary with the nature of the job and the nature of the personnel available. How should this mix be determined? It appeared that these decisions were being made very poorly, or even avoided, in part due to the difficulties of predicting the impact of alternative decisions.

A particularly interesting issue in this project was the apparent lack of users for the trade-off analysis tool. It seemed that no one had the responsibility to make this type of trade-off because no one knew how to do it. This situation was due, in part, to a perceived inadequacy of the scientific and technical knowledge upon which to base methods and tools. Further, and perhaps more important, organizational divisions of responsibility typically result in trade-off issues crossing "turf" boundaries. Thus, training and aiding are usually designed independent of each other.

The payoff sought by designing a trade-off analysis tool was the ability to integrate the analysis of training and aiding requirements. This would enable determination of mixes of training and aiding that potentially would result in improved performance and safety with reduced costs.

NATURALIST PHASE FOR INTELLIGENT COCKPIT

Two other naturalists joined me in this effort—Ren Curry (an aeronautical engineer and private pilot) and Norm Geddes (an aeronautical engineer and former Navy fighter pilot).

Starting Point

We all entered this effort having much aviation-related experience. Other colleagues and I had been exploring the concept of an intelligent cockpit in R&D for over 10 years (e.g., Hammer and Rouse, 1982; Rouse, Rouse, and Hammer, 1982). We were also thoroughly knowledgeable of the R&D literature in this area, to which we were regular contributors. While this starting point would seem to provide obvious justification for skipping the naturalist phase, fortunately we did not make that mistake. Otherwise, we would have not fully understood the nuances of factors likely to influence pilots' acceptance of the intelligent cockpit.

The Process

We interviewed 10 fighter pilots. We asked them how they felt about advanced technology in their cockpit. They said that it is great when it works, but all too often it doesn't—95 percent of the time is not good enough. One pilot said that once a system has given him bad advice three times, he turns it off for good—three strikes and you're out.

The pilots were asked how they felt about automation. They said that it is great for housekeeping tasks such as preflight checks and engine monitoring. They were much less positive about automation for central tasks such as flight control and flight management.

We asked the pilots how they felt about automation intervening and taking control of the aircraft. Nine out of ten said that this would be acceptable if they were unconscious and the aircraft was headed for a certain crash. They were all more positive if they could personally tailor the conditions of intervention to the specific mission and their preferences. In other words, such intervention was more acceptable if they could feel that they were explicitly delegating authority to perform specific tasks in particular situations.

Finally, we asked the pilots about the characteristics of the perfect copilot. They said that the perfect copilot knows exactly what information and help they want and need, provides it, and otherwise does not get in the way.

Results

Clearly, and not surprisingly, pilots want to be in charge. They are not categorically opposed to advanced technology and automation. However, it must be reliably supportive.

Pilots want to delegate authority to the automation, perhaps in preflight, rather than have it predetermined by somebody else, that is, designers. Ideally, the intelligent cockpit would know what pilots want, do it for them, and not get in the way. The challenge, of course, was to produce a concept that pilots would perceive to have the above characteristics (Rouse, Geddes, and Curry, 1987).

NATURALIST PHASE FOR DESIGN INFORMATION SYSTEM

I was joined in this effort by one other naturalist—Bill Cody, a psychologist who had worked for several years for a major aircraft manufacturer.

Starting Point

We had been working on aspects of this problem for about five years prior to this effort beginning (e.g., Rouse and Rouse, 1984; Morehead and Rouse, 1985; Rouse, 1986). As a result, we were thoroughly familiar with the R&D literature on information systems and design support, as well as frequent contributors to this literature.

Three previous experiences on related efforts had taught us several important lessons (Rouse and Cody, 1989b). In one effort, the customer's perceptions of the users' needs were a bit off target. Consequently, the resulting product (Frey et al., 1984; Rouse, 1984) was not as useful as we had anticipated. This was due to the fact that some aspects of the product supported decisions that users were typically organizationally prohibited from making.

In another effort, both we and the customer defined the user population much too narrowly. This small population liked the system we produced (Frey and Wiederholt, 1986; Hunt and Frey, 1987), and they continue to use it. However, a few of our design choices were inappropriate for the potentially much larger population—for example, we made an unfortunate choice for a host workstation.

In the third effort, the customer's perceptions of the users' needs were way off. We performed a small naturalist study to prove this. We won the battle but lost the war—we never got beyond the first phase of this project. This risk should be noted. It is a risk that I feel has to be occasionally taken.

I have repeatedly distinguished between the customer and the user in the above discussion and occasionally earlier. This distinction between the purchaser of a product or system and the eventual user is very important. In Chapter 4, this issue is elaborated within a broader framework that deals with the topic of who influences the success of a design effort.

Based on the above three experiences, we were ready, willing, and enthusiastic about pursuing a full-blown naturalist phase. The goal was to study in depth the nature of design and designers' wants and needs with regard to design support.

The Process

We used questionnaires, interviews, and an observational study. The population queried, using one or more of these methods, included 240 "designers" across disciplines (engineering, human factors, and training), across organizations (industry and government), and across phases of system development, ranging from R&D to full-scale development (Rouse, Cody, and Boff, 1990).

Four questionnaires were distributed at professional meetings. Although the details of the four questionnaires differed somewhat, the basic contents and administration procedures were the same. Each effort solicited information on six main topics:

- Respondents' background and experience,
- Characteristics of their organizational contexts,
- Typical design issues and information needed to resolve them,
- Sources of this information that respondents valued, and why,
- Information bottlenecks that they experienced, and
- Reactions to suggested forms of design support that might overcome these bottlenecks.

Three sets of interviews were performed (Rouse and Cody, 1988; Cody, 1989; Rouse and Cody, 1989a). Interviewees were identified through personal acquaintance and by referral from acquaintances. Each interviewee had participated in the design or test and evaluation of at least one major operational system. Most had extensive experience in advanced development projects as well.

The structured interview format we used concentrated on six topics:

- Interviewee's background and work experience,
- Time allocation among design activities,
- Example design problems and activities,
- Opinions about the strengths and weaknesses of current system design practices,
- Suggested changes to practices, and
- Reactions to suggested forms of design support.

The observational study involved taking up residence with members of a design group over a six-month period while they pursued the early stages of designing a new aircraft cockpit. Design "diaries," interviews, and observations of design integration meetings were used to identify the problems these designers faced, questions they asked, information they sought, and answers they produced.

Results

Not surprisingly, we found that design groups or teams are central to designing complex systems. These teams are usually multidisciplinary.

Designers spend their time in both group and individual activities. For journeymen and seasoned designers, the time allocation is typically 30 percent in group activities and 70 percent in individual activities. Junior designers spend more time in group activity for the purpose of learning. Very senior designers spend more time in group activity, serving as coaches and mentors.

The design group or team has several roles. The group is usually involved with decomposing the statement of work or other descriptions of objectives, requirements, and specifications. Based on this decomposition, the group will set technical goals, as well as allocations of person-hours and schedule, for members of the group. Pursuit of these technical goals is predominantly an individual activity. The group subsequently reviews the results of these individual efforts.

The organization, both of the company and the marketplace, strongly affects both group and individual activities. Company policies and procedures directly influence activities. Success criteria and reward mechanisms, both internal and external to the company, affect motives and values. Corporate and market cultures influence, for example, relative weightings on performance, cost, and quality.

Designers' activities as they seek information and solve problems can be described in the context of levels of abstraction and aggregation (Rasmussen, 1986, 1988). The concept of level of abstraction can be used to discriminate among design issues concerned with the purpose, function, or form of a product. The aggregation dimension discriminates among design tasks occurring at the system, subsystem, assembly, and component levels. By characterizing designers' tasks in terms of the abstraction–aggregation space, powerful concepts of design support can be envisioned. We return to these concepts in Chapter 4.

The methods whereby designers perform their tasks can be described in several ways. The Work Breakdown Structure method is a common approach to organizing project-oriented activities in both industry (Blanchard, 1981) and government (Coutinho, 1977). Design projects also typically have key milestones such as the System Requirements Review, Preliminary Design Review, and Critical Design Review.

Design task performance can also be described in terms of methods of synthesis and analysis. Computer-based drawing and prototyping tools are increasingly being used. Simulation tools have become more accessible and less expensive. The extent to which analytical methods are used varies with disciplines, although all disciplines rely heavily upon the intuitive use of experience and expertise.

A common denominator among all tasks and methods of performance is the access and utilization of different types of information. This information includes that which is generated by the design project. Examples are statements of work, block diagrams, flowcharts, and minutes of group meetings.

Much information external to the design project is also accessed. This includes information on science and technology in general, as well as specific information on particular technologies (e.g., components). External information also encompasses that which is gained from users, customers, and the marketplace.

The overall perspective that emerges is one of designers heterarchically moving among numerous tasks, accessing and utilizing a wide variety of information sources. This process is strongly influenced by group processes and shaped by the organizational environment. The challenge was to develop a design information system concept that comprehensively supports this overall process in an integrated manner. In Chapter 4, we consider how this challenge was approached.

NATURALIST PHASE FOR TRADE-OFF ANALYSIS TOOL

I was the only naturalist involved in this effort. As noted earlier, it is in general much better to have two or more observers in this process. However, circumstances did not allow this.

Starting Point

I had been working for several years developing computational models (Rouse, 1985, 1987b, 1990a) for analyzing trade-offs among human resource investments. For example, is it better to select highly skilled employees, train people extensively, or design products and systems so that they can be used by people with minimal skills and training? Of course, the best answer is probably some combination of the three approaches, but which combination?

Somewhat in parallel, I have over the years served on a variety of government advisory committees that have, slowly, provided me with the bigger picture associated with larger-scale trade-offs. For instance, the military services have until recently completely separated the design of a new system from consideration of how people will be trained to operate and maintain the system. Consequently, trainability problems can emerge when it is too late to change the system design in order to resolve them.

Finally, I was reasonably knowledgeable of the R&D literature in this area. However, I anticipated some surprises once I cast the net wider in the naturalist phase.

The Process

I interviewed about 20 experts across government and industry. This included the military services, government contractors, automobile manufacturers, and insurance companies.

Most of the interviews were by telephone. The few in-person interviews were quite lengthy, typically involving multiple visits.

My questions focused on the nature of training vs. aiding trade-offs, people's perceptions of the importance of these trade-offs, and how their agency or company addressed them. I also tried to gain an understanding of likely avenues for changing the process.

Results

The consensus was that the ability to make training vs. aiding trade-offs would be of tremendous value. However, the difficulty of acquiring the data to support rigorous, quantitative trade-offs was often noted.

Interviewees felt that such trade-offs are done poorly now, or not done at all, because of the lack of tools and, more importantly, organizational boundaries. Consequently, the market for this tool would have to created. Nevertheless, they perceived that the market was there to be created.

A frequent concern was the level of knowledge and skills that one could reasonably assume on the part of likely users of this tool. There is a need for the tool to be accessible and usable by novices, journeymen, and experts. Further, each type of user should perceive "value added" even if they do not fully utilize the tool. The challenge in this project, therefore, was not only developing a method to perform trade-offs, but also devising the packaging of this method to allow a broad range of users (Rouse, 1990b; Rouse and Cacioppo, 1989).

TRANSITIONING TO THE MARKETING PHASE

Once one has embraced the notion of a naturalist phase, there is a tendency to feel that one never knows enough. It seems that more and more study is needed before one can comfortably proceed.

However, the naturalist phase does not have to be "finished" in order to initiate the marketing phase. One only needs to know enough about users, organizations, and the environment to be able to proceed with two activities. First, one should be able to formulate hypotheses about users, organizations, and the environment that are testable in the marketing phase. Second, one should be in a position to develop initial product/system concepts for which one wants to get initial measurements of viability, acceptability, and validity.

REFERENCES

Blanchard, B. S. (1981). *Logistics engineering and management.* Englewood Cliffs, NJ: Prentice-Hall.

Cody, W. J. (1989). Designers as users: Design supports based on crew system design practices. *Proceedings of the American Helicopter Society 45th Annual Forum,* Boston.

Coutinho, J. P. (1977). *Advanced system development management.* New York: Wiley.

Frey, P. R., Sides, W. H., Hunt, R. M., and Rouse, W. B. (1984). *Computer-generated display system guidelines.* Volume 1: *Display design.* (Rept. NP-3701, Vol. 1). Palo Alto, CA: Electric Power Research Institute.

Frey, P. R., and Wiederholt, B. J. (1986). KADD—An environment for interactive knowledge-aided display design. *Proceedings of 1986 IEEE International Conference on Systems, Man, and Cybernetics,* pp. 1472–1477.

Hammer, J. M., and Rouse, W. B. (1982). Design of an intelligent computer-aided cockpit. *Proceedings of 1982 IEEE Conference on Cybernetics and Society,* pp. 449–453.

Hunt, R. M., and Frey, P. R. (1987). Knowledge-aided display design (KADD) system: An evaluation. *Proceedings of 31st Annual Meeting of the Human Factors Society,* pp. 536–540.

Morehead, D. R., and Rouse, W. B. (1985). Computer-aided searching of bibliographic data bases: Online estimation of the value of information. *Information Processing and Management, 21,* 387–399.

Rasmussen, J. (1986). *Information processing and human-machine interaction: An approach to cognitive engineering.* New York: North-Holland.

Rasmussen, J. (1988). A cognitive engineering approach to the modeling of decision making and its organization in process control, emergency management, CAD/CAM, office systems, and library systems. In W. B. Rouse (Ed.), *Advances in man-machine systems research* (Vol. 4, pp. 165–243). Greenwich, CT: JAI Press.

Rouse, S. H., Rouse, W. B., and Hammer, J. M. (1982). Design and evaluation of an onboard computer-based information system for aircraft. *IEEE Transactions on Systems, Man, and Cybernetics, SMC-12,* 451–463.

Rouse, W. B. (1984). *Computer-generated display system guidelines.* Volume 2: *Developing an evaluation plan* (Rept. NP-3701, Vol. 2). Palo Alto, CA: Electric Power Research Institute.

Rouse, W. B. (1985). Optimal allocation of system development resources to reduce and/or tolerate human error. *IEEE Transactions of Systems, Man, and Cybernetics, SMC-15,* 620–630.

Rouse, W. B. (1986). On the value of information in system design: A framework for understanding and aiding designers. *Information Processing and Management, 23,* 217–228.

Rouse, W. B. (1987a). Designers, decision making, and decision support. In W. B. Rouse and K. R. Boff (Eds.), *System design: Behavioral perspectives on designers, tools, and organizations* (pp. 257–283). New York: North-Holland.

Rouse, W. B. (1987b). Model-based evaluation of an integrated support system concept. *Large-Scale Systems, 13,* 33–42.

Rouse, W. B. (1990a). Human resource issues in system design. In N. P. Moray, W. R. Ferrell, and W. B. Rouse (Eds.), *Robotics, control, and society.* London: Taylor & Francis.

Rouse, W. B. (1990b). Training and aiding personnel in complex systems: Alternative approaches and important tradeoffs. In H. R. Booher (Ed.), *MANPRINT: An approach to systems integration.* New York: Van Nostrand Reinhold.

Rouse, W. B., and Cacioppo, G. M. (1989). *Prospects for modeling the impact of human resource investments on economic return.* Washington, DC: Dept. of the Army, Office of the Deputy Chief of Staff for Personnel.

Rouse, W. B., and Cody, W. J. (1988). On the design of man-machine systems: Principles, practices, and prospects. *Automatica, 24,* 227–238.

Rouse, W. B., and Cody, W. J. (1989a). A theory-based approach to supporting design decision making and problem solving. *Information and Decision Technologies, 15,* 291–306.

Rouse, W. B., and Cody, W. J. (1989b). Information systems for design support: An approach for establishing functional requirements. *Information and Decision Technologies, 15,* 281–289.

Rouse, W. B., Cody, W. J., and Boff, K. R. (1990). The human factors of system design: Understanding and enhancing the role of human factors engineering. *Instructional Journal of Human Factors in Manufacturing.*

Rouse, W. B., Geddes, N. D., and Curry, R. E. (1987). An architecture for intelligent interfaces: Outline of an approach to supporting operators of complex systems. *Human-Computer Interaction, 3,* 87–122.

Rouse, W. B., and Rouse, S. H. (1984). Human information seeking and design of information systems. *Information Processing and Management, 20,* 129–138.

Chapter 4

The Marketing Phase

The purpose of the marketing phase is introducing product concepts to potential customers and users. In addition, its purpose includes planning for measurements of viability, acceptability, and validity. Furthermore, initial measurements should be made to test the plans, as opposed to the product, to uncover any problems before proceeding.

It is important to keep in mind that the product and system concepts developed in this phase are primarily for the purpose of addressing viability, acceptability, and validity. Beyond what is sufficient to serve this purpose, minimal engineering effort should be invested in these concepts. Beyond preserving resources, this minimalist approach avoids, or at least lessens, "ego investments" in concepts prior to knowing whether or not the concepts will be perceived to be viable, acceptable, and valid. The phenomenon of premature ego investment is considered again later in this chapter when prototyping is discussed.

These types of problem can also be avoided by pursuing more than one product concept. Potential customers and users can be asked to react to these multiple concepts in terms of whether or not each product concept is perceived as solving an important problem, solving it in an acceptable way, and solving it at a reasonable cost. Each person queried can either react to all concepts, or the population of potential customers and users can be partitioned into multiple groups, with each group only reacting to one concept.

The marketing phase results in an assessment of the relative merits of the multiple product concepts that have emerged up to this point. Also derived is a preview of any particular difficulties that are likely to later emerge. Concepts can be modified, both technically and in terms of presentation and packaging, to decrease the likelihood of these problems. This process is illustrated in the discussion of the case studies later in this chapter.

BUYING INFLUENCES

As noted frequently, the success of a product or system depends on more than convincing users of the validity, acceptability, and viability of the concept. Other types of individuals also strongly influence the success of a product or system. The roles played by these individuals have been characterized by Miller and Heiman (1985). Their book focuses on sales— in this section, their constructs are recast for design.

Economic Influence

The role of this influence is to give final approval for a "sale." Hence, the economic influence is often also called the "customer." For design, a "sale" may constitute funding of an R&D project or approval to proceed with full-scale development.

Typically, there is only one economic influence per sale. This influence may be an individual or a set of people such as a board or committee. This individual or group has direct access to funds, releases funds, has discretionary use of funds, and veto power over use of funds.

The focus of the economic influence is on the "bottom line" and impact on the organization. A typical question is, "What kind of return will we get on this investment?"

For design, the answer to this question is not always in monetary terms. For example, the development of new capabilities with advanced technology is often a central element of the bottom line. A new high-end product may be an important component of the organization's marketing plan, even though the product is produced at a loss (e.g., racing cars produced by automobile manufacturers).

The economic influence is not always a customer in the usual sense of the term. For design decisions, particularly in large organizations, the economic influence may be internal to one's own organization.

From this perspective, all designers from program managers to project engineers should be aware of who controls the resources necessary to their design efforts—in other words, be aware of the customer in their efforts. Thus, everyone in a design organization can be customer-oriented. However, everybody does not necessarily have the same customer.

User Influence

In light of previous discussions, it should not be surprising that the role of this influence is to make judgments about the impact of the product or system on job performance. There are often several or many user influences. It is not uncommon for there to be large differences of opinions among users.

Users are typically the people who will operate, maintain, or manage use of the product or system. Accordingly, they will have to live with the results of the design effort. As emphasized repeatedly, there is a direct link between users' success and the success of the product or system.

The focus of the user influence is on the job to be done. A typical question is, "How will it work for me?" This question can occur on at least two levels.

On the surface, this question concerns job performance. Will work be easier, faster, more productive? Will rewards reflect these improvements?

On a deeper level, this question concerns the impact of the product or system on users' roles. The necessity of change and its likely impact can create uncertainties for users. Anticipating the implications of these issues, as well as dealing with them, is discussed at length later in this chapter.

As with economic influences, or customers, it is quite common for user influences to reside in one's own organization, especially in large organizations. Just as every designer has customers (i.e., controllers of resources one wants), every designer has users—those who use the results of one's efforts. Therefore, every designer has one or more people who can be queried about viability, acceptability, and validity.

Technical Influence

The role of this influence is to screen out proposals, particularly because of perceived validity and evaluation problems. There are often several or many technical influences. It is not unusual for two or more technical influences to have very different opinions on a single technical issue.

When this happens, it is important to smooth out these differences if possible, since they can undermine a design effort, or at least distract it.

Technical influences judge the quantifiable aspects of proposed design efforts. They also judge intermediate results, such as presented at system requirements reviews and preliminary design reviews.

Technical influences serve as gatekeepers, making recommendations. They seldom can say "yes" and give final approval for a project. However, they can say "no" and often do.

The focus of technical influences is on the product or system. Typical questions are, "Will it solve the problem?" and "Will it satisfy requirements?" The first question concerns validity, and the second evaluation.

Technical influences are often staff specialists in the customer's organization. For design projects within large design organizations, these influences are probably technical specialists from other functions or departments within one's own organization.

The Coach

The role of the coach is to act as a guide for the "sale" of the design effort or project at hand. I have found it to be essential to develop at least one coach for each project. For large organizations, it is likely that coaches may change as one's proposal advances. In such situations, a good coach will help to line up the next coach.

Coaches can be found in the buying organization, in one's own organization, or outside of both. People who have recently retired from the buying organization can occasionally be good coaches and, more often, quite helpful in identifying coaches. Friends and relatives are occasionally helpful, but should not be the sole avenue for identifying coaches. Undoubtedly the best way to find coaches is by "networking" via professional colleagues and acquaintances.

Good coaches tend to be highly experienced and very savvy, at least relative to the niche within which one is operating. Coaches provide and interpret information about the current situation, the identity and preferences of the other buying influences, and the ways in which all influences can mutually benefit.

The focus of a coach is on this particular proposed effort. A good coach wants *this* proposal to win. Thus, a typical questions is, "How can we pull this off?"

Buying Influences vs. Measurements

Figure 4.1 summarizes the role of the four buying influences. Simply put, the economic influence or customer is concerned with viability, users are focused on acceptability, and technical influences emphasize validity. Further, each of the influences tends to influence the other.

Coaches are clearly the key to the whole process. Coaches facilitate relationships and guide understanding. Also of great importance is that they serve as cheerleaders.

METHODS AND TOOLS FOR MEASUREMENT

How does one measure the perceptions of the buying influences relative to the viability, acceptability, and validity of alternative product and system concepts? Figure 4.2 lists the appropriate methods and tools for answering this question, as well as their advantages and disadvantages.

	VIABILITY	ACCEPTABILITY	VALIDITY
ECONOMIC	●	◒	◒
USER	◒	●	◒
TECHNICAL	◒	◒	●
COACH	✪	✪	✪

● PRIMARY

◒ SECONDARY

✪ FACILITATING

Figure 4.1. Influences vs. measurements.

METHODS AND TOOLS	PURPOSE	ADVANTAGES	DISADVANTAGES
Questionnaires	Query large number of people regarding preferences for product's functions.	Large population can be inexpensively queried.	Low return rates and shallow nature of responses.
Interviews	In-depth querying of small number of people regarding reactions to and likely use of product's functions.	Face-to-face contact allows in-depth exploration of nature and perceptions of product's functions and benefits.	Difficulty of gaining access, as well as time required to schedule and conduct.
Scenarios	Provide feeling for using product in terms of how functions would likely be used.	Inexpensive approach to providing rich impression of product's functions and benefits.	Written scenarios are not as compelling as visual presentation and require users' willingness to read.
Mock-ups	Provide visual look and feel of product.	Strong visual image can be created and reinforced with photographs.	Necessarily emphasize surface features which are not always product's strength.
Prototypes	Provide ability to use product, typically in a fairly limited sense.	Very powerful and compelling approach to involving potential users.	Relatively expensive and not fully portable; sometimes lead to inflated expectations.

Figure 4.2. Methods and tools for measurement.

Questionnaires

This method can be used to obtain the reactions of a large number of people to alternative functions and features of a product or system concept. Typically, people are asked to rate the desirability and perceived feasibility of functions and features using, for example, scales of 1 to 10. Alternatively, people can be asked to rank order functions and features.

As noted when questionnaires were discussed in Chapter 3, low return rate can be a problem. Moreover, one typically cannot have respondents clarify their answers, unless telephone or in-person follow-ups are pursued. This tends to be quite difficult when the sample population is large.

I have found that questionnaires can present problems if they are the only methods employed in the marketing phase. The difficulty is that responses may not discriminate among functions and features. For example, respondees may rate as 10 the desirability of all functions and features.

This sounds great—one has discovered exactly what people want! However, another interpretation is that the alternatives were not sufficiently understood for people to perceive different levels of desirability among the choices. Asking people to rank order items can eliminate this problem, at least on the surface. However, my perception is that questionnaires usually are not sufficiently rich to provide people with real feelings for the functionality of the product or system.

Interviews

Interviews are a good way to follow up questionnaires, perhaps for a subset of the population sampled if the sample was large. As noted earlier, questionnaires are a good precursor to interviews in that they cause interviewees to have organized their thoughts before the interviews. In-person interviews are more useful than telephone interviews because it is much easier to iteratively uncover perceptions and preferences during face-to-face interaction.

Interviews are a good means for determining people's a priori perceptions of the functionality envisioned for the product or system. It is useful to assess these a priori perceptions independent of the perceptions that one may subsequently attempt to create. This assessment is important because it can provide an early warning of any natural tendencies of potential customers and users to perceive things in ways other than is intended in the new product or system. If problems are apparent, one may decide to change the presentation or packaging of the product to avoid misperceptions.

Scenarios

At some point, one has to move beyond the list of words and phrases that describe the functions and features envisioned for the product or system. An interesting way to move in this direction is by using stories or scenarios that embody the functionality of interest and depicts how these functions might be utilized.

These stories and scenarios can be accompanied by a questionnaire within which respondents are asked to rate the realism of the depiction. Further, they can be asked to explicitly consider, and perhaps rate, the validity, acceptability, and viability of the product functionality illustrated. It is not necessary, however, to explicitly use the words "validity," "acceptability," and "viability" in the questionnaire. Words should be chosen that are appropriate for the domain being studied—for example, viability may be an issue of cost in some domains and not in others.

It is very useful to follow up these questionnaires with interviews to clarify respondents' comments and ratings. Often the explanations and clarifications are more interesting and valuable than the ratings. An example of this is noted in the later discussion of the marketing phase for the design information system.

Mock-ups

Mock-ups are particularly useful when the form and appearance of a product or system are central to customers' and users' perceptions. For products such as automobiles and furniture, form and appearance are obviously central. However, mock-ups can also be useful for products and systems where appearance does not seem to be crucial.

For example, computer-based systems often tend to look quite similar— they frequently look like computer terminals or workstations. The only degree of freedom is what is on the display. One can exploit this degree of freedom by producing mock-ups of displays using photographs or even viewgraphs for use with an overhead projector.

One word of caution, however. Even such low-budget presentations can produce lasting impressions. One should make sure that the impression created is such that one wants it to last. Otherwise, as noted repeatedly, there may not be an opportunity to make a second impression.

Prototypes

Prototypes are a very popular approach and, depending on the level of functionality provided, can give potential customers and users hands-on experience with the product or system. For computer-based products, rapid prototyping methods and tools have become quite popular because they enable the creation of a functioning prototype in a matter of hours.

Thus, prototyping has two important advantages. Prototypes can be created rapidly and enable hands-on interaction. With these advantages, though, come two important disadvantages.

One disadvantage is the tendency to produce ad hoc prototypes, typically with the motivation of having something to show potential customers and users. It is very important that the purpose of the prototype be kept in mind. It is a device with which to obtain initial measurements of validity, acceptability, and viability. Consequently, one should make sure that the functions and features depicted are those for which these measurements are needed. One should not, therefore, put something on a display simply because it seems like a good idea or looks nifty. This can be a difficult impulse to avoid.

The second disadvantage is the tendency to become attached to one's prototypes. At first, a prototype is merely a device for measurement, to be discarded after the appropriate measurements are made. However, once the prototype is operational, there is a tendency for people, including the creators of the prototype, to begin to think that the prototype is actually very close to what the final product or system should be like. In such situations, it is common to hear someone say, "Maybe with just a few small changes here and there ... "

Prototypes can be very important. But one must keep their purpose in mind and avoid "rabid" prototyping! Also, care must be taken to avoid premature ego investments in prototypes. The framework for design presented in this book can provide the means for avoiding these pitfalls.

Summary

During the naturalist phase, the goal was to listen. In the marketing phase, one can move beyond just listening. Various methods and tools can be used to test hypotheses that emerged from the naturalist phase, and obtain potential customers' and users' reactions to initial product and system concepts.

Beyond presenting hypotheses and concepts, one also obtains initial measurements of validity, acceptability, and viability. These measurements are in terms of quantitative ratings and rankings of functions and features, as well as more free flow comments and dialogue. For the latter, the primary skill needed is, again, listening.

MARKETING PHASE FOR INTELLIGENT COCKPIT

The results of the naturalist phase for the intelligent cockpit were quite clear. Pilots want to be in charge, although they are not categorically against advanced technology as long as it is reliably supportive. They want to be able to explicitly delegate authority to automation, rather than having it predetermined during design. Ideally, they said, an intelligent cockpit would know what pilots want, do it for them, and not get in the way.

As this project moved into concept development, there was serious disagreement within the design team about how to proceed. One faction wanted to focus on adding intelligence to existing and emerging avionics software architectures. The other faction, which we were part of, wanted to be totally pilot-centered and not constrained by existing concepts.

The customer's program manager made the Solomon-like decision to split the group into two teams. Each team had one month to develop a concept consistent with their respective philosophies. During this period, there was to be no communication between the two teams.

The Squire Concept

Some of this time was used to step back and ask what it means to be supportive in the ways that pilots wanted. Analogies were considered such as Sherlock Holmes' partner Dr. Watson, Don Quixote's sidekick Sancho Panza, and Lord Peter Wimsey's manservant Bunter. A bit of reading about these characters led to the observation that knight's squires are like what pilots said they wanted.

A few pilots were asked what they thought of this notion and, as expected, they very much liked this image. Such a positive response from potential users is great. However, it also struck us that the image would probably be viewed as too old-fashioned by many buying influences.

This led to the notion of putting R2D2 of *Star Wars* fame on the back of a horse behind the knight. This anachronistic image was created as the first viewgraph for the presentation of the concept. The difficulty, of course, was what the rest of the viewgraphs should look like—what functions should squire R2D2 provide?

After much thinking and discussion, it was concluded that three high-level functions were necessary. First, the pilot needed help in managing the likely information explosion in the cockpit. Second, there was a need to avoid the consequences of likely pilot errors in the increasingly complex cockpit environment. Third, a way of flexibly utilizing automation was required to provide pilots with assistance as needed, but also keep pilots "in charge" and in the control loop.

These three functions were elaborated and embedded in a concept called an intelligent interface (Rouse, Geddes, and Curry, 1987; Rouse, Geddes, and Hammer, 1990). The primary modules of the intelligent interface are defined in Figure 4.3. Note that the term "user" is employed in this figure because the intelligent interface concept has been shown to be applicable to other domains such as process control and manufacturing—pilots can be viewed as one type of user.

The two additional modules shown in this figure, intent interpretation and user model, are necessary for the three primary modules to work. The necessity and functioning of all the modules are explained in Chapter 8. They are defined here because they were explicitly considered in the study that is discussed in the next section.

The Squire concept, including the knight and R2D2 on horseback as well as five block diagrams showing how it would all work, was presented at the end of the one-month competition. Pilots responded very positively and management endorsed our proposal. The pilot-centered concept became the centerpiece of the project.

It is important to note, however, that the other team did not lose. Subsequently, both concepts were integrated, with some modifications, into an overall pilot-centered package.

It is interesting to mention that the name "Squire" did not last very long. It captured people's attention and provided a compelling metaphor. But it was quickly decided that it could not stand on its own without the great risk of seeming *very* outdated. Nevertheless, the Squire concept had served an important purpose.

An Initial Study

With the above endorsement, we began to elaborate the concept, with special emphasis on technology feasibility and development. Once the details of the functioning of the system were worked out, and it became clear that it was definitely feasible, an initial assessment of validity, ac-

MODULE	PURPOSE/FUNCTIONS
Interface Manager	The role of the interface manager is to manage the flow of information to the user, as well as requests for information from the user, so as to utilize effectively the user interface (i.e., displays and controls), as well as the user's information processing resources and input/output channels.
Error Monitor	The role of the error monitor is to decrease the frequency of user errors to the extent that such an approach does not prohibit valuable user activities, as well as monitor user behaviors and provide feedback that enables quick detection of anomalies and reversal or compensation for consequences in a timely manner.
Adaptive Aiding	The role of adaptive aiding is to modify its support to meet current user needs and capabilities in a way that utilizes human and computer resources optimally, while also assuring that users retain a maximum degree of control but are not overwhelmed when the magnitude or nature of demands exceeds human abilities to perform acceptably.
Intent Interpretation	The purpose of intent interpretation is to utilize user's actions, as well as the states of the world, system, and user, to infer activities, awareness, and intentions in the context of the goals, plans, and scripts appropriate for the domain of interest.
User Model	The purpose of the user model is to estimate current and predicted user state in terms of activities, awareness, intentions, resources, and performance; the model includes components for intent interpretation (see above), resource requirements, and performance predictions.

Figure 4.3. Modules within intelligent interface.

ceptability, and viability was pursued (Sewell, Geddes, and Rouse, 1987).

Six former fighter pilots, who were currently commercial pilots, were the population queried. First, the design of the intelligent interface was explained to each of them in terms of the five modules in Figure 4.3.

Each pilot was then asked to rate each module on a seven-point scale ranging from –3 to +3, where zero indicated neutral or indifferent, positive numbers a perceived "success," and negative numbers a perceived "failure." Four ratings were made, one each for validity, acceptability, and viability, and the fourth for personal desirability.

After these initial ratings, pilots were talked through a scenario, where particular events were emphasized, and the intelligent interface was not present. Four ratings were made for each event.

The scenario and events were then repeated, illustrating what the intelligent interface would be doing during each event. Four ratings were again made for each event. Beyond ratings, pilots' verbal comments and responses to other questions were tape-recorded throughout each session.

Analysis of the resulting data showed that pilots rated the intelligent interface as highly successful in terms of validity and viability. However, they only rated it moderately successful relative to acceptability and personal desirability, with the latter somewhat higher than the former.

Specifically, pilots were very positive about the interface manager, but apprehensive about adaptive aiding and intent interpretation. They said it was not always clear who was in charge. It is important to note that this was an acceptability concern, rather than a validity or viability issue.

From these results, it was concluded that the intelligent interface concept was sound. However, it would have to be packaged and introduced quite carefully to assure pilots that they are in charge.

Use of Prototypes

The transition of this project from the marketing phase to the engineering phase, as well as its continued development in the engineering phase, involved many prototypes. These prototypes were regularly shown to all of the buying influences (i.e., economic, user, technical, and coaches) to assess their reactions, as well as maintain a common vision.

As of this writing, the intelligent cockpit is being readied for integration into both an upgrade of an existing aircraft and a new aircraft. The key issue has become getting this complex software system to run in real time on standard avionics processors.

We return to the intelligent interface in Chapter 8. The functioning of this concept is discussed in more detail, and the broader implications of the concept are considered.

MARKETING PHASE FOR DESIGN INFORMATION SYSTEM

The results of the naturalist phase for the design information system indicated that designers needed support as they moved heterarchically (i.e., other than hierarchically) among numerous tasks, accessing and utilizing a wide variety of information sources. Further, this support would have to fit in with the group processes that are central to design and the organizational environment within which design occurs.

Use of Questionnaires and Interviews

Four questionnaire surveys were performed to assess designers' preferences for a variety of design support functions (Rouse, Cody, and Boff, 1990). While the results were interesting, and certainly inexpensive to obtain, they suffered from the aforementioned difficulty with questionnaires. Specifically, all proposed functionality was highly desired.

Three sets of interviews were performed (Rouse and Cody, 1988; Cody, 1989; Rouse and Cody, 1989). These were much richer in that one could ask designers to clarify and explain their preferences. From these efforts emerged a cogent hypothesis about the nature of design and potentially appropriate design support functionality.

This hypothesis, which was mentioned in Chapter 3, was that information access and utilization occur in the context of a set of archetypical design tasks that can be described in a two-dimensional design space (Rouse and Cody, 1989). One dimension of this space concerns level of abstraction. A system can be described in terms of its purpose, function, or form. The other dimension concerns level of aggregation—whether information is sought and utilized on the system, subsystem, assembly, or component level.

The archetypical tasks and the design space are discussed in more detail in Chapter 8, where approaches to design of aiding systems are considered. This level of detail is not needed to discuss the role of these constructs in the marketing phase.

Use of Scenarios

The design space was used as a basis for characterizing design scenarios. It was hypothesized that sequences of tasks, or trajectories, in the design space could be used to generate realistic design stories.

A 15-page story was generated involving 20 task episodes in the design space. The story was written in the first person and involved redesign of an aircraft cockpit because of what is eventually proven to be a design-induced human error problem. Embedded in this story were the design support functions listed in Figure 4.4.

Six aircraft crew system designers were asked to read the story and answer a variety of questions about the story's realism and relevance. They were also asked to rate each of the functions in Figure 4.4 relative to perceived value (i.e., validity and acceptability) and feasibility (i.e., viability). Subsequent to filling out the questionnaires and rating forms, they were interviewed and asked to clarify and explain their responses.

Results showed that they placed differential value on the functions depicted in the story. For example, support for technical administration

1. Support browsing and linking on-line sources of design information.
2. Retrieve text and graphics from several sources (e.g., past design documents, statements of work, requests for proposals, standards and regulations, handbooks, term glossaries, design guides).
3. Retrieve and explain models of system and human behavior and performance.
4. Support creation, editing, and execution of qualitative and numerical simulations of human and system performance.
5. Provide tutoring in specialized domains and on the use of specific design tools and models.
6. Support technical administration (e.g., keep track of design decisions, check design parameters against constraints).
7. Support managerial administration (e.g., keep track of calendar, schedule meetings, prevent interruptions).

Figure 4.4 Design support functions.

was highly valued, while support for managerial administration was only moderately valued. Interestingly, perceived feasibility was high for managerial administrative support and moderate for technical administration support. Designers' ratings of feasibility were negatively affected, for the most part, by perceptions that the costs of developing the necessary databases and knowledge bases would be immense.

Designers were asked whether or not the 15-page scenario captured a portion of their jobs. They said that the problem attacked was representative, but the overall computer-based approach was not. However, they said that they wished they did their jobs this way. We had not thought to ask this question in this manner. Fortunately, the interviews brought this out.

It is useful to note another type of scenario that was used in this project to attempt to capture the broader context within which the design information system would likely exist. A script was developed for a video involving four people pursuing a design problem and occasionally interacting with the design information system. The video was shot in one day and, as might be expected, was very rough. Nonetheless, it was quite useful in shaping the project group's vision of a possible future.

Use of Prototypes

Prototypes also served as a means for shaping and maintaining the group's vision of the design information system. Two very different prototypes were developed, more than a year apart, to illustrate how the range of functionality shown in Figure 4.4 might be integrated and packaged.

These prototypes were never taken to the point of development that would have been necessary to enable users to interact with them. However, reactions from economic and technical buying influences were obtained and were instrumental in avoiding premature elaboration of concepts that would not have been subsequently endorsed.

We return to the design information system in Chapter 8 where it is used as an example of a method for performing requirements analyses for aiding systems.

MARKETING PHASE FOR TRADE-OFF ANALYSIS TOOL

The results of the naturalist phase for the trade-off analysis tool indicated a real need for rigorous methods of formulating and resolving training vs. aiding trade-offs. The reasons for the absence of an existing tool included a lack of methodology and technology, organizational structures that typically create boundaries across which such trade-offs are difficult to make, and a rather heterogeneous potential user population.

Before attempting to develop a concept for a trade-off analysis tool, a wide range of alternatives were explored from the human factors, engineering, and management literature (Rouse and Johnson, 1989). It was concluded that no single approach was adequate, but several composite approaches looked promising.

A "shallow" prototype was developed that embodied these approaches—it was shallow in the sense that it was only a series of displays of steps in the process, without any computational depth. Intrinsic to this prototype was an example problem involving incorporation of a head-up display in long-haul trucks and the subsequent analysis of training vs. aiding trade-offs.

This prototype was shown to economic and technical buying influences—recall that there were no existing users for the tool. These buying influences felt that too much was being expected of potential users. It had been assumed that users of the tool would have outlined the training and aiding alternatives, as well as the primary trade-offs, prior to using the tool.

Based on the comments and suggestions of the people who looked at the prototype, the concept was substantially broadened to support novices, journeymen, and expert users, as well as enable users to identify training and aiding alternatives with the tool (Rouse et al., 1989). This broader concept of the trade-off analysis tool is discussed in Chapter 9.

A new prototype was developed that reflected this broader concept. It was shown to economic and technical influences. They responded quite positively and endorsed proceeding in the direction represented by the new prototype.

Emphasis then shifted to applying the approach to several realistic problems, for example, aircraft mechanics and medical technicians. There were two primary goals. First, it was important to assess the availability of requisite data and other information to support the overall approach to trade-offs. Second, there was a need to have multiple credible examples of use of the tool in order to begin to create the user population (or market niche) that did not yet exist.

We return to the trade-off analysis tool at the end of Chapter 9, where the fundamental trade-offs between training and aiding are discussed. Alternative computational approaches to resolving trade-offs are also considered.

PLANNING FOR USER ACCEPTANCE

Two concerns often underlie user acceptance problems (Rouse and Morris, 1986). First, users may perceive, unfortunately sometimes correctly, that the product or system will not function as promised. This may be due to designers' naîveté about the problem being addressed, or perhaps because of a lack of maturity of the technology being employed.

Users may be apprehensive that, despite this concern, the product or system will be purchased anyway. They will be forced to use it, or at least have it sit in the corner. They will then have to regularly explain to management why it is not used.

The second concern that tends to underlie user acceptance problems is people's perceptions that their roles, jobs, and tasks will change in undesirable ways. To deal with this concern appropriately, it is necessary to understand what people value about their roles, jobs, and tasks.

It appears that a central issue is people's opportunity to exercise desirable levels of discretion over their tasks in terms of skill, judgment, and creativity. Some persons value opportunities to exercise skill most highly—they like the act of performing their tasks. Other's value opportunities

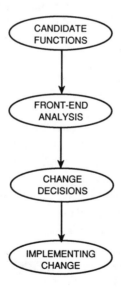

Figure 4.5. Planning for user acceptance (Rouse and Morris, 1986).

to exercise judgment—they see themselves as decision makers. Still, others value opportunities to exercise creativity—they like generating the alternatives. Of course, most people value some combination of these opportunities.

Moreover, they want discretion over some aspects of their jobs and not others. For example, most people want to drive their cars, but not shift the transmission or perform the maintenance.

Finally, most people do *not* want complete discretion. They want structure and boundaries created within which they have much discretion, and outside of which they do not have responsibilities.

The above considerations led to the development of a three-step process for planning for user acceptance that is depicted in Figure 4.5. One begins this process with a set of candidate functions for the product or system. These candidate functions have typically emerged from an analysis of how to meet performance, productivity, and/or safety objectives. The question at this point is whether or not user acceptance problems are likely, and how these problems can be avoided.

Figure 4.6*a* provides four guidelines for *front-end analysis.* The purpose of this list is determination of the basis for potential user acceptance problems. Upon completing this analysis, one should know the functions for which impact on discretion is the central concern. One then must decide whether or not changing these functions is appropriate.

Figure 4.6*b* provides four guidelines for *change decisions.* This list describes conditions necessary for people to accept the changes envisioned. These conditions are necessary but not sufficient. It is also important to consider how people will anticipate changes for which the decision to proceed has been made. Figure 4.6*c* provides guidelines for planning and *implementing changes.*

It is important to emphasize that the structured set of guidelines presented here are focused on anticipating and ameliorating unwarranted perceptions by users of negative impacts of changes on their discretion. Other issues such as, for example, ease of use will affect acceptance after the product or system is developed—these types of issues are discussed in Chapters 5 and 7. In the marketing phase, however, the focus is on anticipating problems and planning accordingly.

TRANSITIONING TO THE ENGINEERING PHASE

As the results of the marketing phase emerge, it should become increasingly clear what users want and need, as well as what the other buying influences will be willing to endorse. One now must focus on the process of developing the product or system. In addition, one now must crisply face technological reality.

Since technology feasibility and development will have been proceeding in parallel with planning for validity, acceptability, and viability—as well as making initial measurements—the engineering phase should not hold any substantial, and unfortunate, surprises relative to technological feasibility. It is still quite possible, though, to encounter schedule and cost problems. From this perspective, one may want to initiate the engineering phase in parallel with the last segments of the marketing phase to foster more interaction than might otherwise occur.

REFERENCES

Cody, W. J. (1989). Designers as users: Design supports based on crew system design practices. *Proceedings of the American Helicopter Society 45th Annual Forum,* Boston.

1. Characterize the functions of interest in terms of whether or not these functions currently require humans to exercise significant levels of skill, judgment, and/or creativity.
2. Determine the extent to which the humans involved with these functions value the opportunities to exercise skill, judgment, and/or creativity.
3. Determine if these desires are due to need to feel in control, achieve self-satisfaction in task performance, or perceptions of potential inadequacies of technology in terms of quality of performance and/or ease of use.
4. If need to be in control or self-satisfaction are *not* the central concerns, determine if the perceived inadequacies of the technology are well founded. If so, eliminate the functions in question from the candidate set; if not, provide demonstrations or other information to familiarize personnel with the actual capabilities of the technology.

Figure 4.6a. Front-end Analysis.

5. To the extent possible, only change the system functions that personnel in the system feel should be changed (e.g., those for which they are willing to lose discretion).

6. To the extent necessary, particularly if number 5 cannot be followed, consider increasing the level and number of functions for which personnel are responsible so that they will be willing to change the functions of concern (e.g., expand the scope of their discretion).

7. Assure that the level and number of functions allocated to each person or type of personnel form a coherent set of responsibilities, with an overall level of discretion consistent with the abilities and inclinations of the personnel.

8. Avoid changing functions when the anticipated level of performance is likely to result in regular intervention on the part of the personnel involved (e.g., assure that discretion once delegated need not be reassumed).

Figure 4.6b. Change Decisions.

9. Assure that all personnel involved are aware of the goals of the effort and what their roles will be after the change.

10. Provide training that assists personnel in gaining any newly required abilities to exercise skill, judgment, and/or creativity and helps them to internalize the personal value of having these abilities.

11. Involve personnel in planning and implementing the changes from both a systemwide and individual perspective, with particular emphasis on making the implementation process minimally disruptive.

12. Assure that personnel understand both the abilities and limitations of the new technology, know how to monitor and intervene appropriately, and retain clear feelings of still being responsible for system operations.

Figure 4.6c. Implementing change.

Miller, R. B., and Heiman, S. E. (1985). *Strategic selling.* New York: William Morrow.

Rouse, W. B., and Cody, W. J. (1988). On the design of man-machine systems: Principles, practices, and prospects. *Automatica, 24,* 227–238.

Rouse, W. B., and Cody, W. J. (1989). A theory-based approach to supporting design decision making and problem solving. *Information and Decision Technologies, 15,* 291–306.

Rouse, W. B., Cody, W. J., and Boff, K. R. (1990). The human factors of system design: Understanding and enhancing the role of human factors engineering. *International Journal of Human Factors in Manufacturing.*

Rouse, W. B., Frey, P. R., Wiederholt, B. J., and Zenyuh, J. P. (1989). *The TRAIDOFF concept.* Norcross, GA: Search Technology, Inc.

Rouse, W. B., Geddes, N. D., and Curry, R. E. (1987). An architecture for intelligent interfaces: Outline of an approach to supporting operators of complex systems. *Human-Computer Interaction, 3,* 87–122.

Rouse, W. B., Geddes, N. D., and Hammer, J. M. (1990). Computer-aided fighter pilots. *IEEE Spectrum, 27,* 38–41.

Rouse, W. B., and Johnson, W. B. (1989). *Computational approaches for analyzing tradeoffs between training and aiding.* Brooks Air Force Base, TX: Air Force Human Resources Laboratory.

Rouse, W. B., and Morris, N. M. (1986). Understanding and enhancing user acceptance of computer technology. *IEEE Transactions on Systems, Man, and Cybernetics, SMC-16,* 965–973.

Sewell, D. R., Geddes, N. D., and Rouse, W. B. (1987). Initial evaluation of an intelligent interface for operators of complex systems. In G. Salvendy (Ed.), *Cognitive engineering in the design of human-computer interaction and expert systems.* New York: North-Holland.

Chapter **5**

The Engineering Phase

The purpose of the engineering phase is developing a final design of the product or system. Much of the effort in this phase involves using various design methods and tools in the process of evolving a conceptual design into a final design. In addition to synthesis of a final design, planning and execution of measurements associated with evaluation, demonstration, verification, and testing are pursued.

It is important to keep in mind the inherent conflict between design and evaluation that was discussed in Chapter 2. Figure 5.1 illustrates this conflict. Design begins by considering the requirements for an effective design, proceeds to being concerned with the understandability of human–system communication, and finally focuses on the compatibility of displays, input devices, and the environment.

Evaluation, in contrast, inevitably must first deal with compatibility. If a user cannot see the displays or reach the keyboard, then understandability and effectiveness cannot be addressed. Similarly, if a user can see the displays but labels, symbols, and so on, are meaningless, then effectiveness is not addressable.

These two perspectives create the greatest conflict during the engineering phase, when designers, who are struggling with effectiveness issues, must deal with evaluators' complaints when they cannot get by compatibility problems. Designers will sometimes say, "Don't bother me with

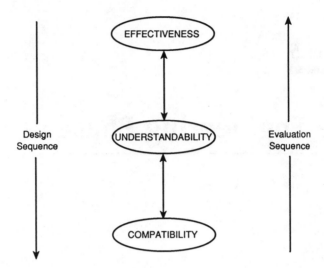

Figure 5.1. Top-down design vs. bottom-up evaluation (Rouse, 1984, 1986a).

such nits—just evaluate the concept." Evaluators will respond, "Because of your nits, I can't see the concept."

I do not think that this tension can be eliminated. However, if everyone realizes that such conflicts are inevitable, conflicts can be more quickly clarified and resolved.

FOUR-STEP DESIGN PROCESS

In this section, a four-step process for directing the activities of the engineering phase and documenting the results of these activities is discussed (Hunt and Maddox, 1986). The essence of this process is a structured approach to producing the four documents, which is indicated in Figure 5.2.

A structured approach to producing design documentation can be invaluable, even if the overall design process during the engineering phase tends to be very unstructured occasionally (Parnas and Clements, 1986). Documentation produced in this manner can be especially valuable for tracing back from design decisions to the requirements and objectives that motivated the decisions. For example, I have found that suggested design

STEP	DOCUMENT
1. Identify product or system goals, functions, and objectives	System Objectives Document (SOD)
2. Identify specific information and control requirements	System Requirements Document (SRD)
3. Develop conceptual design of product or system	Conceptual Design Document (CDD)
4. Develop final design of product or system	System Design Document (SDD)

Figure 5.2. Four-step design process (Hunt and Maddox, 1986).

changes are much easier to evaluate and integrate into an existing design when one can efficiently determine why the existing design is as it is.

Case Studies

The four steps of the design process are illustrated using two case studies. The first case study involved designing an upgrade to the physical security system for a large manufacturing complex. This customer used a crew of nine security guards to control and monitor ingress and egress of many thousands of workers each day. The stated goal, at least initially, was to give each crew member a computer-based system to help them to do their jobs. Throughout the remainder of this section, this first case study is referred to as the "security system."

The second case study involved designing an operator console for control of a satellite communications system. This case study is referred to as the "operator console."

System Objectives Document

The first step in the process is developing the system objectives document (SOD). The SOD contains three attributes of the product or system to be designed: goals, functions, and objectives.

Goals are characteristics of the product or system that designers, users, and other buying influences would like the product or system to have. Goals are often philosophical choices, frequently very qualitative. There are usually multiple ways of achieving goals. Goals are particularly useful for providing guidance for later choices.

Example goals for the security system included

- Automate functions as much as possible,
- Minimize paper documentation generated,
- Make all work areas accessible to handicapped individuals, and
- Provide personnel with meaningful, professionally rewarding jobs and tasks.

Example goals for the operator console included

- Ensure security of data bases,
- Minimize amount of keyboard entry,
- Use operator pacing of all system activities,
- Maximize use of direct display manipulation, and
- Accommodate knowledge and skills of particular labor categories.

Functions define *what* the product or system should do, but not *how* it should be done. Consequently, there are usually alternative ways to provide each function. The definition of functions subsequently leads to analysis of requirements.

Example functions for the security system included

- Access control,
- Communications,
- Alarm control, and
- Crisis management.

Example functions for the operator console included

- Alignment,
- Link control,
- Mission replay, and
- Dynamic simulation.

Objectives define the activities that must be accomplished by the product or system in order to provide functions. Each function has at least one, and often five to ten, objectives associated with it. Objectives are typically phrased as imperative sentences beginning with a verb.

As an illustration from the security system, the function termed "alarm control" had the following associated objectives:

- Detect, identify, and classify alarm occurrences,
- Allow any alarm to be functionally tested,
- Detect and identify alarm malfunctions,
- Allow alarms to be bypassed centrally, and
- Determine which alarms are bypassed and why.

For the operator console, the following objectives were associated with the function "alignment":

- Allow locations of surface equipment to be entered,
- Instruct operators how to collect location information,
- Verify alignment data measurements as necessary, and
- Calculate correction factor for hardware delays.

There are two purposes for writing a formal document listing goals, functions, and objectives. First, as noted earlier, written documents provide an audit trail from initial analyses to the "as-built" product or system. The system objectives document provides the foundation for all subsequent documents in the audit trail for the engineering phase. The second purpose of the SOD is that it provides the framework—in fact, the outline—for the system requirements document.

All stakeholders should be involved in the development of the SOD. This includes at least one representative from each type of user group, as well as representatives of the other buying influences. This is important because the SOD defines what the eventual product or system will and will not do. All subsequent development assumes that the functions and objectives in the SOD form a necessary and complete set.

The contents of the SOD can be based on interviews with subject matter experts, including operators, maintainers, managers, and trainers. Baseline and analogous systems can also be valuable, particularly for determining objectives that have proven to be necessary for providing specific functions.

Much of the needed information will have emerged from the marketing phase. At the very least, one should have learned from the marketing phase what questions to ask and who to ask. All the stakeholders in the process should have been identified and their views and preferences assessed.

The level of detail in the SOD should be such that generality is emphasized and specifics are avoided. The activities and resulting document should concentrate on what is desirable. Discussion of constraints should be delayed—budgets, schedules, people, and technology can later be considered.

System Requirements Document

Once all the stakeholders agree that the SOD accurately describes the desired functions and objectives for the product or system, the next step is to develop the system requirements document (SRD). The purpose of the SRD is to identify all information and control requirements associated with each objective in the SOD. The intent of the analysis is to determine what information the product or system must supply to the user, and what actions the user must be capable of taking with the system.

The information and control requirements analysis should be very detailed. To illustrate, one of the objectives for the operator console was "display system performance data to the operator." This is much too general a statement of system requirements to be useful in detailed system design.

A much more detailed accounting of the information requirements for this objective is necessary before one can specify system components to meet this objective. Answers to the following questions can provide the necessary detail for this example:

- What data define system performance?
- What are the characteristics of each data item?
- How do data requirements change with operational phase?

Frey and his colleagues (1984) have developed a way of characterizing data items that can be mapped to choices of display elements, an issue that is addressed in final design. Answers to the following questions are used to characterize each data item:

- Is it alphanumeric, pictorial, and so forth?
- What is the range, if numeric?
- How rapidly must it be read?
- To what precision must it be read?
- Is trend/historical information required?
- Is it used qualitatively or quantitatively?
- Is it compared to other values or setpoints?

Answers to these questions can be obtained from potential users, including operators, maintainers, managers, and trainers.

For evolutionary designs, baseline and analogous systems can be studied to find answers to these questions. However, if the product or system being designed has no antecedent, subject matter expertise can be very difficult to find. In this case, answers to the above questions have to come from engineering analysis and, if necessary, validated empirically.

The product of the analyses outlined in this section is the system requirements document. The SOD serves as the outline for the SRD. Objectives from the SOD are used as headings in the SRD. Information and control requirements associated with each objective are tabulated with minimal descriptive prose.

As with the SOD, the SRD should be reviewed and approved by all stakeholders in the design effort. This approval should occur before beginning development of the conceptual design.

Beyond its use as input to the conceptual design, the SRD can also be very useful for determining the functional significance of future design changes. In fact, the SRD is often used to answer questions that arise concerning why particular system features exist at all. Without this link, there is a tendency when resources (e.g., display space and processing speed) are scarce to delete features when no one can remember why these features were included.

Conceptual Design Document

The conceptual design of a product or system should accommodate *all* information and control requirements as parsimoniously as feasible within the state of the art. The conceptual design, as embodied in the conceptual design document (CDD), is the first step in defining *how* the final system will meet the requirements of the SRD. The CDD is a working document that evolves into the final design for the product or system.

The CDD should describe a complete, workable system that meets all design objectives. While many alternatives will surely emerge in the process of devising a conceptual design, it is usually best if the CDD does not provide alternatives. The reason for this is that the CDD should be the means whereby a consensus is reached.

Realistically, one should expect considerable disagreement as the conceptual design evolves. However, the CDD should not reflect these disagreements. Instead, the CDD should be iteratively revised until a consensus is reached. At that point, all stakeholders should agree that, if no constraints existed, the conceptual design documented in the CDD is a desirable and appropriate product or system.

Constraints should be avoided in order to creatively define the ideal system. This ideal can then be subjected to various cost/benefit analyses. Even if constraints eventually eliminate several aspects of this ideal, it can then serve as the defining document for future investment opportunities.

We have found it useful to capture competing visions for the conceptual design using a notion called *evolutionary architectures,* which is summarized in Figure 5.3. The important distinction underlying this construct is the tension between "blue-sky" innovative thinking and "down-to-earth" practical thinking.

Blue-sky thinking can be risky, but down-to-earth thinking can be boring and noncompetitive. The buying influences usually will endorse some concept between these extremes. The solution, conceptually at least, is to propose level B with level A as the "fallback," and level C as the "vision"—perhaps what is called a "preplanned product improvement."

- Level A: What you know you can do.
- Level B: What you are willing to promise.
- Level C: What you would like to do.
- Principle: Conceptual architecture should be capable of
                potentially supporting all three levels.

Figure 5.3. Evolutionary architectures.

The principle of evolutionary architectures is that the core architecture of the system should be compatible with all three levels, so that falling back and/or future improvements do not involve restarting. In a sense, adopting this principle means that one is designing two or more systems, or a family of systems, at the same time. It can be difficult to convince buying influences that such an investment will pay off. However, customers, users, and particularly investors who have a reasonable planning horizon will often see the merits of this principle—this issue is further discussed in Chapter 10.

System Design Document

The fourth and final step in the design process involves synthesizing a detailed design. Associated with the detailed design is the system design document (SDD). This document describes the "production" version of the product or system, including block diagrams, engineering drawings, parts lists, and manufacturing processes.

The SDD links elements of the detailed design to the functionality within the CDD, which is in turn linked to the information and control requirements in the SRD, which are in turn linked to the objectives within the SOD. These linkages provide powerful means for efficiently revising the design when, as is quite often the case, one or more buying influences do not like the implications of their earlier choices. With the audit trail provided by the four-step design process, evaluating and integrating changes are much more straightforward. As a result, good changes are readily and appropriately incorporated, and bad changes are expeditiously rejected.

Summary

While it probably could go without saying, it is important to note that the four-step, document-oriented process described in this section adds some up-front "overhead"to the design process. I find quite often that people question the wisdom of investing in this process. Several experiences illustrate the value of the process and, therefore, why the overhead is worthwhile.

My colleagues and I were asked by a customer to evaluate a computer-generated display system for use in nuclear power plant control rooms. At our first meeting, we told this customer that the system could not be evaluated without knowing the objectives it was designed to achieve—we needed their equivalent of the SOD. They did not have such documentation and went so far as to ask us what we thought would be good objectives for the system.

We refused this request and they then went away for one month to prepare an objectives document. When we met one month later, they proudly gave us their objectives document, which was quite good. They also noted that, in the process of clarifying and documenting their objectives, they had come to several realizations that caused them to change the system substantially prior to our evaluating it. Many person-hours and much time was lost because objectives were not considered in the first place.

In the aforementioned design effort for the security system, it was noted that the customer initially requested that we develop computer-based workstations for each of the nine members of the security crew per shift. We asked if nine computers in front of nine people was the primary goal of the project.

After much discussion and debate, it was agreed that access control was the goal and the nine workstations were one possible solution. This allowed us to break free of the nine-person concept—the final design was a three-person crew. This obviously could result in very substantial labor savings, the return on which would dwarf the investment necessary to follow the four-step process.

Development of the design information system discussed in Chapters 3 and 4 involved detailed analyses of requirements and the system functionality implied—this analysis is discussed in Chapter 8. The results of these analyses were over 600 instances of needed functionality, which were grouped into a relatively small number of functional categories.

As this information was being used to develop a conceptual design, a wide variety of intuitively appealing changes was suggested by two colleagues. I found that the possibility of revising the concept to satisfy these suggestions was helped tremendously by being able, in this case via computer, to trace all design decisions back through information and control requirements, to objectives, functions, and goals. As a result, common ground with these stakeholders was reached in a few hours, with several of their suggestions readily incorporated and a few rejected. Without the linkages provided by the four-step design process, it would have been almost impossible to revise several hundred functions consistently and comprehensively.

DESIGN ISSUES

The third and fourth steps of the design process—conceptual and final design—involve formulating and resolving many design issues. A large fraction of these issues involve hardware and software alternatives and problems. Product geometry, system dynamics, software data structures, power requirements, real-time computation constraints, and display generation rates are examples of these types of issues. These issues cross many disciplines within engineering and computer science and cannot be addressed within the scope of this book. It is important, however, to consider those issues that relate specifically to humans interacting with products and systems.

Human–System Interaction Issues

For our purposes, human–system interaction issues can be thought of as being of two types. The first type includes those issues that tend to be very sensitive to design philosophies such as human-centered design. In other words, the way in which these issues are resolved is directly and substantially influenced by adopting a human-centered perspective. Issues within this category include allocation of functions among humans and machines, aiding systems, training systems, and trade-offs between training and aiding. Function allocation is discussed in this section, while the other issues are addressed in Chapters 8 and 9.

The second type of human–system interaction issue includes those issues whose resolution are less directly affected by design philosophy. Workspace design is such an issue. Of concern here is that reach, motion, and sight requirements are compatible with humans' abilities and limitations. The effects of environmental stresses such as noise or vibration are also of concern. Workstation design is also in this second category of issues. Display layout and format, choice of symbology and labels, and use of windowing and paging are examples of issues in this class.

This second type of human–system interaction issue is important, and successful resolution is necessary to human-centered design, especially for compatibility and understandability. However, human factors issues at this level are not the central "value added" of the framework and methods outlined in this book. These issues can and should be appropriately resolved following good human factors principles, independent of whether or not the notions in this book are adopted.

Of the many books dealing with this second type of human–system interaction issue, those by Moraal and Kraiss (1981), Bailey (1982), Kantowitz and Sorkin (1983), and Meister (1985) are particularly useful because of their design orientation. In addition, Boff and Lincoln's (1988) *Engineering Data Compendium* provides a rich source of compiled and well-formatted data on human performance and perception. These data primarily relate to compatibility issues and, to a limited extent, understandability.

Rouse's (1986b) categorization of human–system interaction issues provides a basis for putting the two types of issues outlined above in perspective. Further, this categorization serves to clarify in detail the relative priorities of human-centered design in general and this book in particular. Succinctly, human–system interaction can be characterized on four levels:

- Environmental compatibility,
- Input–output requirements,
- Information processing, and
- Authority/responsibility.

Environmental Compatibility. At this level, one is concerned with ensuring that humans can survive and function successfully within the human–system environment. Short- and long-term occupational safety and health issues are paramount. Design of life support and protection systems such as needed in outer space or deep-sea operations has environmental compatibility as a primary concern.

Input–Output Requirements. At the next highest level, one is interested in ensuring that the input–output requirements for human–system interaction are compatible with humans' abilities and limitations. Issues at this level are concerned with sensing, or input, primarily via vision and hearing, but also using touch, smell, taste, and kinesthetic sense. Similarly, one is concerned with affecting, or output, via speech and motion. The anthropometric characteristics of humans both enable and constrain the range of motions, and forces, that are possible and desirable.

Beyond sensing and affecting, other factors intrinsic to humans, i.e., aptitude, ability, style, attitude, and motivation can affect their performance. The environment also includes factors that can substantially affect humans' abilities to perform. Examples include noise, illumination, temperature, vibration, motion, acceleration, weightlessness, and biochemical and radiological hazards.

Input–output requirements influence the appropriate nature of displays and controls—the means for sensing inputs and affecting outputs, respectively. Display technology has advanced rapidly in recent years, and issues associated with computer-generated displays have received considerable attention. Structured approaches for designing such displays include the aforementioned guidelines by Frey and his colleagues (Frey et al., 1984), as well as a more recent computer-based version of this approach that provides on-line advice as displays are being designed (Frey and Wiederholt, 1986; Hunt and Frey, 1987). This structured approach to display design is currently being augmented to provide a more thorough treatment of graphic and pictorial displays (Sewell, Rouse, and Johnson, 1989).

Not only has computer technology provided a much richer set of ways to display information, but it also has resulted in a dramatic increase of the range and types of information that can be displayed. As a result, all of the relevant and useful information for a particular application seldom fits on a single display "page." Often there are many pages, typically with hierarchical relationships among them. Unfortunately, the hierarchical, multipage displays that result can be quite complex, with even modest levels of depth leading to degraded human performance (Henneman, 1988). Considerable effort is now being devoted to studying such "hypermedia" systems (Glushko, 1989).

Information Processing. A primary emphasis within human–system design is the process whereby inputs are transformed into outputs by humans. Issues at this level include memory, attention, and workload, the latter of which is discussed later in this chapter. Information processing is also influenced by task requirements which are often characterized in terms of activities such as monitoring, control, decision making, and problem solving. These types of activities are also discussed later in this chapter and in subsequent chapters.

Authority/Responsibility. The highest-level issue within human–system interaction concerns questions such as who is in control, how authority reflects responsibility, and how design characteristics should support this balance. These types of questions lead to the fundamental human–system interaction issue of function allocation.

Allocation of Functions

There is no issue more central to human-centered design than the alloca-tion of functions. The process of function allocation is how one decides which functions will be performed by humans, which will be performed by machines, and when. This process has become more challenging in recent years owing to advances in information technology that have resulted in machines with capabilities similar to some human intellectual abilities. Consequently, one is free to explore novel allocation schemes in which responsibility for system functions can be assigned to either of two equally capable entities (i.e., human or computer).

In general, there tends to be little disagreement about what constitutes a desirable outcome of the function allocation process. One would like to determine the allocation that best accomplishes system performance ob-jectives and also creates an acceptable job for the humans involved. Sev-eral prescriptions for making allocation decisions have appeared in the design literature. Bailey (1982) categorizes these into three groups.

Comparison allocation approaches (Doring, 1976; Meister, 1971) re-commend that one carefully analyze the skill requirements and perfor-mance criteria for each system function. One then consults lists that com-pare the relative abilities of humans and machines for activities that bear some resemblance to the function under consideration (e.g., Fitts, 1951). The allocation decision is a simple matter of selecting the superior per-former.

In *leftover allocation* (Chapanis, 1970), the strategy is to automate as many system functions as technology will permit, and assign the leftover duties to humans. This approach assumes that the leftover functions will form coherent and satisfactory jobs for humans.

Economic allocation methods emphasize the costs associated with de-veloping and operating the system. For each function, the problem is to determine whether it is more economical to select, train, and pay a person to perform the function, or to design, develop, and maintain equipment to perform the same function.

Unfortunately, despite technological advances, agreement about design goals, and the availability of several allocation techniques, function alloca-tion remains a difficult problem (Air Force Studies Board, 1982; Price, 1985). There appear to be three reasons for this.

First, these methods are typically portrayed as one of the earlier steps in an idealized design process. Once functional requirements are defined, one then applies one of the above strategies to determine a proper split of responsibilities between humans and technology. The underlying assumption is that once allocation is settled, integration and detailed design can proceed without further concern for allocation.

This view is unrealistic because the success of allocation decisions depends on how they are implemented. One cannot know whether or not an allocation is good or bad without designing the system to a sufficient level of detail to be able to evaluate both human and machine performance. Hence, allocation, design, and evaluation tend to occur repeatedly, with each pass through the three steps considering fewer alternatives and producing greater design detail.

A second limitation of the above approaches to allocation is that they implicitly assume that human behavior is best understood in terms of a limited-capacity, serial information processor that is capable of handling one task at a time. As a result, functions are treated one at a time and jobs for humans are designed that consist of collections of supposedly independent tasks. But performance of simultaneous tasks yields performance that is often less than the sum of the parts, and occasionally more than the sum. In other words, tasks often compete with each other and sometimes complement each other.

The third limitation with current allocation methods is the implied goal of partitioning system functions into two mutually exclusive sets, one set for humans and one for machines (typically computers). This goal does not take full advantage of the overlap in intelligent capabilities between hu-

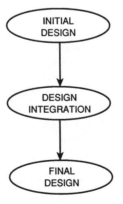

Figure 5.4. Three steps of function allocation.

mans and machines that advances in artificial intelligence and adaptive control have created. Intelligent machines enable allocation decisions to be situation-dependent for those functions that both humans and machines can perform acceptably. Hence, it is feasible to design an allocation scheme that assigns functions dynamically according to which performer is more capable at the moment to absorb the imposed demands.

The above three considerations led us to develop an iterative approach to allocation involving the three steps shown in Figure 5.4. Each of the three steps involves the three activities shown in Figure 5.5. The remainder of this section provides an overview of this approach (Rouse, 1985; Rouse and Cody, 1986).

The input to this process is a "function timeline" which describes functional requirements. These requirements usually vary in time, both in terms of whether or not each function is required at all, and in terms of changing conditions, criteria for acceptable performance, and criticality to success.

Use of the methodology results in two products. One is an "allocation timeline" which specifies, for each time period, the allocation of each function to either human or machine. To the extent that the allocation is dynamic, the timeline specifies the most likely allocation.

For each function that will be performed by humans for at least one time period, use of the methodology also produces a task design that completely describes the resulting task in terms of displays, input devices, and operating procedures. Also documented are the human performance models and data used in the course of designing the task.

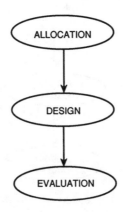

Figure 5.5. Activities within each step of function allocation.

Initial Design. For this step, the objective is to develop an approximate but comprehensive scoping of the total allocation problem. The function timeline is sorted by function to produce a list of times when each function is demanded. More importantly, this process reveals the types of condition, as well as ranges of performance criteria and criticalities, associated with each function.

An initial allocation is made using comparative allocation procedures such as those mentioned earlier, which employ lists of relative human and machine abilities. Functions are allocated to humans, machines, or potentially both. Functions allocated to humans or both are then converted to tasks by designing initial displays, input devices, and procedures.

For these initial task designs, estimates of human performance and workload are used to assess whether or not humans could perform these tasks acceptably, independent of the demands of other tasks. It is important to emphasize that performance and workload are *predicted* at this point rather than *assessed.* Methods for generating such predictions are discussed later in this chapter.

If performance and workload projections are acceptable, the initial allocation remains unchanged. For marginally acceptable predictions, the initial allocation is not changed, but the functions involved are flagged for future scrutiny. For completely unacceptable predictions, reallocation to machines is considered.

Design Integration. The first pass through the allocation–design–evaluation cycle yields a baseline design that can serve as a rudimentary benchmark. This first pass focuses on single-task performance and workload at different points in time. This is clearly an approximation of reality at best. Thus, the second pass—design integration—emphasizes relationships among multiple tasks at similar points in time.

The primary goal of design integration is to take advantage of complementary relationships among tasks to enable integrated displays, input devices, and/or procedures that enhance performance and reduce workload. For tasks that are conflicting, displays, input devices, and/or procedures are redesigned to separate tasks in terms of time, or modalities and formats of displays and input devices.

Evaluation is performed via both predictions and assessments of performance and workload. Predictions are replaced by empirical assessments when task designs are sufficiently novel that little modeling experience is available. Either type of estimate is compared to criteria values and the allocation modified accordingly.

Final Design. This third and final pass through the allocation–design–evaluation sequence begins by reviewing all allocation decisions. Decisions that were based on marginally acceptable or unacceptable human performance/workload are resolved. Final allocation decisions are heavily influenced by whether the overall automation philosophy emphasizes defaulting to human or machine control.

Tasks that could clearly be performed by either humans or machines are considered for dynamic or time-varying allocation. Approaches to determining when to shift the allocation are first explored. If a feasible approach is identified, issues associated with human–computer interaction are then addressed. Of special importance are the means whereby humans will monitor the performance of those tasks that are dynamically allocated. Dynamic allocation is discussed in detail in Chapter 8.

Evaluation during this final iteration usually requires much more elaborate methods than used earlier. This is due to both the comprehensive nature of the design at this point and the need to provide definitive data to assess achievement of design objectives. Consequently, part-task and full-scope simulators are often used for evaluations at this point. Approaches to evaluation are considered in detail in Chapter 7.

Summary. The function allocation methodology outlined in this section has two central characteristics that enable avoiding the shortcomings of previous approaches. First, the activities of allocation, design, and evaluation are viewed as interdependent. Second, the three steps enable an initial focus on single-task baselines, intermediate consideration of complementary and conflicting tasks, and final emphasis on the interaction and integration of all redesigned tasks.

The referenced reports (Rouse, 1985; Rouse and Cody, 1986) provide flowcharts and procedural descriptions for pursuing each activity and step. These are, however, but one way of achieving the characteristics just described. The important point is that any approach to function allocation should have these characteristics.

Estimation of performance and workload was noted repeatedly in the above discussion. The next subsection provides an overview of approaches to workload. The subsequent main section presents a broad discussion of performance modeling for the purposes of function allocation in particular and design in general.

Approaches to Workload

When allocating functions to humans, one does not only want to know if they can perform acceptably. One also wants to know if they can sustain performance over a duty cycle, a pay period, or a career. In many situations, people will assure that performance is acceptable even if the "costs" of performing are very high.

Workload is a concept for capturing the costs humans pay when they perform. In general, but not always, if workload is too high, performance will eventually suffer, either because people become fatigued or because they are simply unwilling to put up with the high workload. In contrast, if workload is too low, people are easily distracted and become bored.

Thus, there must be some optimal level of workload—not too high and not too low. The search for this optimum level has been the holy grail within a segment of the human factors and experimental psychology community for many years. A variety of perspectives have emerged (Moray, 1979; Gopher and Donchin, 1986).

Approaches to assessing workload include

- Primary task performance—workload must be high when performance cannot be sustained.
- Secondary task performance—workload on primary tasks must be high when performance on secondary tasks degrades.
- Physiological correlates—physiological indices (e.g., eye pupil diameter) change to reflect high workload.
- Subjective reports—reported experiences of high workload are indicative of high workload.

Boff and Lincoln (1988, entry 7.701) review the sensitivity, diagnosticity, and practicality of these approaches to workload assessment. Wierwille and his colleagues (1985) consider the sensitivity and intrusiveness of a variety of approaches.

While these approaches to *assessment* can be useful, they are not sufficient for design where *prediction* is necessary. One would like to know the workload impacts of design alternatives without having to fabricate a design and perform an empirical assessment. We have found two approaches to prediction to be particularly useful, and a third to be very promising.

Workload is too high when a human does not have enough time to do all the things he or she is supposed to. Therefore, at least at extremes, time-oriented measures of workload can be very useful (Moray, 1979). This approach has been quite common in, for example, the aviation and manufacturing industries.

To illustrate, we developed a simulation model of pilots' task performance for a new flight management concept that we were exploring (Walden and Rouse, 1978; Chu and Rouse, 1979). Using queueing theory (i.e., models of behavior in waiting lines) to represent the "serving" of tasks, the fraction of time that pilots were busy was predicted as a function of various design parameters. These predictions were very highly correlated with pilots' subjective assessments of their workload in subsequent experiments.

Another approach that we have found useful is multiple resource theory (Wickens, 1984; Wickens, Tsang, and Pierce, 1985). This theory views humans as having varying information processing resources from perceptual input through cognitive processes through motor output. Information may be input in one form, processed in another form, and output in yet another. The resources drawn upon can vary as a function of external context and internal availability.

Input resources are defined by crossing input modality with processing code. Four combinations result: visual–spatial, visual–verbal, auditory–spatial, and auditory–verbal. Central processing resources are considered to be spatial or verbal. Output takes manual or vocal forms.

This multidimensional characterization of resources provides a useful accounting heuristic for human workload. This heuristic can be used in the function allocation method outlined earlier (Rouse, 1985; Rouse and Cody, 1986). It was embedded within the intelligent cockpit to predict the impacts of task demands on pilot workload—this is further discussed in Chapter 8.

With increasing automation in many complex systems, human workload is no longer driven by activities, multidimensional or otherwise. In these situations, people spend much time monitoring, occasionally they intervene, and only rarely do they exercise complete manual control. While such situations are seldom good examples of human-centered design, they are nevertheless increasingly prevalent.

What influences workload when there is relatively little explicit activity? In two recent studies using process control simulations, it was found that people's perceptions of the state of their current situation strongly relate to their subjective assessments of their workload, which in these studies was expressed in terms of effort required (Morris and Rouse, 1988).

As shown in Figure 5.6, perceptions that the situation has become unacceptable (e.g., excessive oscillation) result in rating increases—the heavy arrows indicate that this transition is the predominant event leading to rating increases. Inadequate explanations of the change, as well as inadequate plans for dealing with the change, result in further rating increases. If the situation becomes again acceptable without intervention, or an explanation is found, or a plan is devised, ratings then decrease. Throughout this process, there may or may not be any human activity. Nonetheless, perceived workload may be varying considerably.

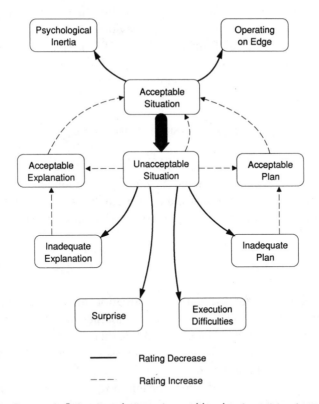

Figure 5.6. Factors influencing changes in workload ratings (Morris and Rouse, 1988).

There are four other constructs shown in Figure 5.6:

- Psychological inertia—increased workload due to, for example, just having had a "close miss" such as nearly losing control of the system.
- Operating on edge—increased workload due to trying to squeeze the last ounce of performance out of the system.
- Surprise—increased workload due to, for example, an event occurring that was not thought to be possible.
- Execution difficulties—increased workload due to, for example, data entry errors.

These constructs provide explanations of changes in workload ratings that were not necessarily correlated with the evolving situation.

These results have provided a basis for a new approach to predicting workload. This approach is based on the notion that people's perceptions of their own workload influence their behaviors which, in some cases, also influence the resulting performance (Rouse, 1990). This approach is yet to be tested but is a promising way of dealing with situations where activity levels are too low to use time-oriented or multiple resource approaches.

MODELING HUMAN–MACHINE INTERACTION

In the discussion of function allocation, the need to predict the impacts of alternative task designs was noted. Several types of models can be used to make such predictions. Models and their predictions also have many other uses within system design. In this section, a wide range of models and uses is discussed.

Types of Models

Models are means for generating information about phenomena (e.g., predictions of their behaviors) by approximating the characteristics and processes underlying the phenomena. While there are many ways to model phenomena, they all fall within three classes or types of models: experiential, empirical, and analytical.

An *experiential model* of a new system might be the previous version of that system (i.e., a baseline) and a characterization of how the new system will be different. An experiential model provides answers to "what if" questions by assuming that the new system's behavior and performance will be very similar to that of the old system except for upgrades envisioned to overcome past deficiencies or provide new capabilities. The assumption that the past is a good predictor of the future is a very good premise for everyday life. It is not, however, as universally applicable in design, where often, but not always, one is creating the future.

An *empirical modeling* effort involves collecting data for conditions and subject populations that are assumed to represent the eventual operating conditions and user population of the system of interest. The behaviors observed and performance measures calculated are useful to the extent that experimental conditions and subject populations are good models of the target application. For this reason, empirical data are not inherently more accurate than, for example, experts' judgments of likely behaviors or performance. In other words, contrary to prevailing wisdom in human factors and experimental psychology, experimental models of human–machine interaction do *not* inherently produce results of higher validity.

Analytical modeling involves constructing a computational representation of the phenomena of interest and computing various characteristics of this representation, typically its response to various manipulations. The phenomena represented may vary in levels of abstraction and aggregation. For instance, a representation might model elemental activities such as keystrokes or more aggregate performance characteristics such as task performance time. As another illustration, a model might represent fairly concrete human–system behaviors such as manual control, or more abstract phenomena such as situation awareness. The most appropriate level of representation depends on the nature of the question being asked and the form of the answer being sought.

Purposes of Models

There are, in general, four major uses of models of human–machine interaction. First of all, the modeling process itself is beneficial. Developing a model requires a very organized and thorough pursuit of all the issues surrounding the phenomena of interest. This is especially true if one pursues modeling to the stage of creating a computer program that incorporates the resulting model. The iterative process of model development and testing provides many insights into the phenomena being studied.

A second use of models is to provide succinct descriptions of data. While a tabulation of performance vs. time may summarize the results of an evaluation, a set of equations or rules are much more useful, as well as more compact. This is particularly true for design decision making.

Models are also useful for designing experiments in general and evaluations in particular. It is not unusual for complex phenomena to have many attributes, often too many to permit comprehensive evaluation. A model of the phenomena can be very useful as a guide for choosing which attributes to emphasize. A sensitivity analysis can be performed with the model to determine which parameters within the model most affect the model's outputs. The parameters whose variations have the most effect can then be chosen for empirical evaluation.

A fourth use of models is to make quantitative predictions. These are particularly useful in system design to answer "what if" questions. Predictive models can also be embedded in system designs as a means of predicting likely human performance and modifying the displays, nature of aiding, and so forth. This idea is discussed in detail in Chapter 8.

Analogies and Constrained Optimality

Choosing an approach to modeling is tantamount to choosing a form of representation for the phenomena of interest. Analogies provide a useful basis for making this choice.

For example, human behavior and performance are sometimes analogous to the behavior and performance of servomechanisms, that is, error-nulling mechanisms such as thermostats. As another illustration, human behavior and performance are often analogous to that of a time-shared computer in the ways that tasks are sequenced and scheduled. If either of these analogies is adopted as the basis for a model, then one has available powerful mathematical and computational machinery for analyzing the model and producing predictions of performance. Later in this section, a wide range of analogies, and associated mathematical and computational methods, are discussed.

The mathematical and computational methods are used to predict the model's outputs (i.e., behavior and performance) as a function of its inputs (i.e., task demands). The calculation or computation process often requires that one assume performance criteria that the behavior of the model is to optimize. It is usually assumed that people adopt the criteria imposed by the task, as opposed to criteria of their own, and attempt to perform as well as possible within their physical and psychological constraints. This notion of constrained optimality enables making strong predictions that, in many cases, are quite accurate. Use of this notion also enables iterative discovery of constraining mechanisms as one attempts to uncover why predictions and observations do not agree.

The use of analogies as a means for choosing representations and associated mathematical and computational methods, and the use of constrained optimality as a criterion for calculation and computation, are very powerful ways for approaching modeling and performance prediction. A companion issue concerns how human–system tasks are conceptualized within an overall model—this conceptualization influences what aspects of the human–system are modeled and how they are modeled.

Control Paradigms

The predominant paradigm for many years was *manual control* (Sheridan and Ferrell, 1974). People are conceptualized as input–output devices that sense inputs and produce outputs so as to control a vehicle or process to follow a path or remain within targeted values. In this paradigm the human is characterized as a system component.

As automatic control of vehicles and processes has become increasingly prevalent, humans' roles in task performance have become more oriented toward giving directions, choosing targets, and occasionally intervening if unusual events occur. This role is called *supervisory control* in that humans supervise automation, which is allocated direct performance of tasks (Sheridan, 1984; Rasmussen, 1986).

With advances in intelligent systems technology, automation has become increasingly sophisticated and the supervisory/subordinate distinction is often not as clear. For example, in some situations humans may monitor the automation and, in other cases, the automation may monitor the humans—illustrations of this concept are discussed in Chapter 8. The result is a new paradigm, *collaborative control* (Rouse, Geddes, and Curry, 1987), where humans and automation must explicitly cooperate to perform tasks.

The distinctions among these three control paradigms have important implications for how task performance is modeled. In manual control, humans are active system components, and modeling their input–output relationships, as affected by the nature of the task, is usually sufficient.

For supervisory control, the automation has the active role in the sense that it performs tasks in response to humans' directions. As a result, one has to model the performance of both the automation and humans, including the inputs and outputs that they supply each other.

Collaborative control is further complicated by the fact that humans and automation serve as "partners" in task performance. Consequently, they have to share much information, including information about what each intends and what each is aware of. As a result of this need, the goal structures of each entity must be considered—this issue is discussed in the context of intelligent interfaces in Chapter 8.

Alternative Representations

Summarizing the discussion thus far, the choice of how to represent phenomena is influenced by several factors:

- Purpose of the modeling effort—is a computational method needed?
- Possible analogies—what is the nature of the behaviors of interest?
- Computational approach—what are relevant constraints and the implications for computational methods?
- Control paradigm—what aspects of human and machine have to be represented in the overall model?

Answers to these questions influence the choice among the alternative representations listed in Figure 5.7. The nature of the mapping shown in this figure is discussed in great detail in Rouse (1980).

ANALOGY	REPRESENTATION
Ideal Observer	Estimation Theory
Servomechanism	Control Theory
Time-Shared Computer	Queueing Theory
Approximate Reasoner	Fuzzy Set Theory
Knowledge-Based System	Rule-Based Models
Pattern Recognizer	Statistical Models

Figure 5.7. Alternative representations (Rouse, 1980).

The *ideal observer* analogy, as well as the notion of constrained optimality, assumes that humans deal with observed uncertainties in an optimal manner, subject to various behavioral constraints such as noisy perceptual processes and limited memory. Use of this analogy enables utilization of estimation theory which provides means for computing the optimal estimates, within the given constraints, which serve as predictions of human performance in the conditions modeled.

The *servomechanism* analogy assumes that humans are optimal feedback controllers subject to behavioral constraints that include noisy perceptual and motor processes, reaction time delays, and sluggish neuromotor responses. Adoption of this analogy, as well as the notion of constrained optimality, enables use of control theory for computing the optimal control actions, within the given constraints, which serve as predictions of humans' control actions in similar circumstances.

The *time-shared computer* analogy assumes that humans optimally sequence and perform tasks subject to constraints such as limited time, switching times, and perceptions of priorities. Use of this analogy enables one to employ queueing theory, which provides means for computing average task waiting times, average number waiting, and fraction of time busy. As noted earlier in this chapter, the fraction of time busy can be used to predict workload in tasks where time constraints are dominant.

The *approximate reasoner* analogy assumes that humans are logical reasoning machines subject to constraints such as not having crisp knowledge of how things work, what connects to what, and which elements belong to different sets. Adopting of this analogy enables using fuzzy set theory which provides means for computing sequences of problem-solving steps, relative to a noncrisp model of the world, which can serve as predictions of action sequences for tasks such as process control and troubleshooting.

The *knowledge-based system* analogy assumes that humans' knowledge is encoded explicitly in verbal if–then statements rather than implicitly in equations or routines of some sort. Behavioral constraints are characterized in terms of knowledge limitations. Use of this analogy enables adoption of rule-based models which include lists of if–then relationships and rule-processing mechanisms for making inferences using the rule set. The result is a sequence of conclusions, some of which may be actions that can serve as predictions of human behavior in situations similar to that for which the rule base was developed.

The *pattern recognizer* analogy assumes that people perform a direct mapping from displayed or perceived features to conclusions or actions, based on statistical relationships drawn from past experience. Statistical models, including neural net models, can be used when this analogy is appropriate to predict likely responses, hit rates, and related metrics.

The representations listed in Figure 5.7 are "pure" in the sense that a single class of representational forms is assumed, along with a single type of mathematical or computational method. Many complicated task situations require more than one representation, which are then integrated within a larger model or modeling framework or package. Several recent books summarize and evaluate alternative models, approaches, and packages for dealing with more complicated modeling problems (Baron and Kruser, 1990; Elkind et al., 1989; McMillan et al., 1989).

An Example

To illustrate the process of choosing among alternative representations, consider the results of an analysis of the tasks associated with use of a head-up display (HUD) in long-haul trucks to assist with collision avoidance. The intended use of the HUD was to project displayed information on the inside of the windshield so that, despite rain, fog, or snow, the driver could "see" potential obstacles without taking his or her eyes off the road.

Analysis of the HUD concept led to identification of four operations tasks and five maintenance tasks associated with the HUD. The operations tasks included situation interpretation, maneuver selection, execution and monitoring, and operating the HUD itself. The maintenance tasks included test verification, fault isolation, repair decisions, replacing units and boards, and degraded mode assessment.

The mapping of these nine tasks to representations is shown in Figure 5.8. Rouse and Johnson (1989) discuss the details of this analysis and the specific models chosen that embody the representations indicated. For our purposes, three general points should be noted about the mapping in Figure 5.8.

First, no single representation was sufficient for all nine tasks. Second, two pairs of tasks were sufficiently interdependent that they had to be dealt with using single forms of representation. Finally, three of the tasks could be sufficiently proceduralized to handle with quite simple models.

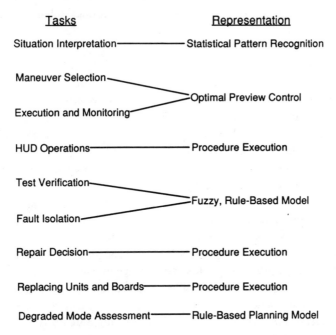

HEAD-UP DISPLAY FOR A LONG-HAUL TRUCK

Figure 5.8. Mapping tasks to representations (Rouse and Johnson,1989).

Summary

Models provide a very powerful means for structuring an analysis, organizing what one knows, and predicting the results of design manipulations. Sometimes fairly accurate quantitative predictions are possible. However, even in those more typical situations where only qualitative predictions are possible, the discipline that modeling brings to a design effort can be invaluable.

TRANSITIONING TO THE SALES AND SERVICE PHASE

The transition to sales and service is a major event. The product or system will have been designed, developed, manufactured, packaged, and shipped, or made ready to be shipped.

It would seem that design should now be done. But, design is not done. There are measurements yet to be made. Further, it is now time to begin the naturalist phase for the next product or product enhancement.

REFERENCES

Air Force Studies Board (1982). *Automation in combat aircraft.* Washington, DC: National Academy Press.

Bailey, R. W. (1982). *Human performance engineering: A guide for system designers.* Englewood Cliffs, NJ: Prentice-Hall.

Baron, S., and Kruser, D. S. (Eds.) (1990). *Human performance modeling.* Washington, DC: National Academy Press.

Boff, K. R., and Lincoln, J. E. (Eds.) (1988). *Engineering data compendium: Human perception and performance.* Wright-Patterson Air Force Base, OH: Air Force Aerospace Medical Research Laboratory.

Chapanis, A. (1970). Human factors in systems engineering. In K. B. DeGreene, (Ed.), *Systems psychology.* New York: McGraw-Hill.

Chu, Y. Y., and Rouse, W. B. (1979). Adaptive allocation of decision making responsibility between human and computer in multi-task situations. *IEEE Transactions on Systems, Man, and Cybernetics, SMC-9,* 769–778.

Doring, B. (1976). Analytical methods in man-machine system development. In K. F. Kraiss and J. Moraal (Eds.), *Introduction to human engineering.* Cologne: TUV Rheinland.

Elkind, J. W., Card, S. K., Hochberg, J., and Huey, B. M. (Eds.) (1989). *Human performance models for computer-aided engineering.* Washington, DC: National Academy Press.

Fitts, P. M. (1951). Engineering psychology and equipment design. In S. S. Stevens (Ed.), *Handbook of experimental psychology.* New York: Wiley.

Frey, P. R., Sides, W. H., Hunt, R. M., and Rouse, W. B. (1984). *Computer-generated display system guidelines.* Volume 1: *Display design.* (Rept. NP-3701, Vol. 1). Palo Alto, CA: Electric Power Research Institute.

Frey, P. R., and Wiederholt, B. J. (1986). KADD—An environment for interactive knowledge aided display design. *Proceedings of 1986 IEEE International Conference on Systems, Man, and Cybernetics,* pp. 1472–1477.

Glushko, R. J. (1989). Hypermedia: Visions of grandeur? *UNIX Review, 8,* 70–80.

Gopher, D., and Donchin, E. (1986). Workload—An examination of the concept. In K. R. Boff, L. Kaufman, and J. P. Thomas (Eds.), *Handbook of perception and human performance* (Chapter 41). New York: Wiley.

Henneman, R. L. (1988). Human problem solving in dynamic environments: Understanding and supporting operators in large-scale complex systems. In W. B. Rouse (Ed.), *Advances in man-machine systems research* (Vol. 4, pp. 121-164). Greenwich, CT: JAI Press.

Hunt, R. M., and Frey, P. R. (1987). Knowledge-aided display design (KADD) system: An evaluation. *Proceedings of 31st Annual Meeting of the Human Factors Society*, pp. 536–540.

Hunt, R. M., and Maddox, M. E. (1986). A practical method for designing human-machine system interfaces. *Proceedings of the 1986 IEEE International Conference on Systems, Man, and Cybernetics*, pp. 407-411.

Kantowitz, B. H., and Sorkin, R. D. (1983). *Human factors: Understanding people-system relationships.* New York: Wiley.

McMillan, G. R., Beevis, D., Salas, E., Strub, M. H., Sutton, R., and van Breda, L. (Eds.) (1989). *Applications of human performance models to system design.* New York: Plenum.

Meister, D. (1971). *Human factors: Theory and practice.* New York: Wiley.

Meister, D. (1985). *Behavioral analysis and measurement methods.* New York: Wiley.

Moraal, J., and Kraiss, K. F. (Eds.), (1981). *Manned systems design: Methods, equipment, and applications.* New York: Plenum.

Moray, N. (Ed.) (1979). *Mental workload: Its theory and measurement.* New York: Plenums

Morris, N. M., and Rouse, W. B. (1988). *Human operator response to error-likely situations in complex engineering systems* (Tech. Rept. 177484). Moffett Field, CA: NASA Ames Research Center.

Parnas, D. L., and Clements, P. C. (1986). A rational design process: How and why to fake it. *IEEE Transactions on Software Engineering, SE-12.*

Price, H. E. (1985). The allocation of functions in systems. *Human Factors, 27,* 33–46.

Rasmussen, J. (1986). *Information processing and human-machine interaction: An approach to cognitive engineering.* New York: North-Holland.

Rouse, W. B. (1980). *Systems engineering models of human-machine interaction.* New York: North-Holland.

Rouse, W. B. (1984). *Computer-generated display system guidelines. Volume 2: Developing an evaluation plan* (Rept. NP-3701, Vol. 2). Palo Alto, CA: Electric Power Research Institute.

Rouse, W. B. (1985). *Computer-aided crew station design: An approach to supporting engineering judgment in function allocation.* Norcross, GA: Search Technology, Inc.

Rouse, W. B. (1986a). Design and evaluation of computer-based decision support systems. In S. J. Andriole (Ed.), *Microcomputer decision support systems* (Chapter 11). Wellesley, MA: QED Information Systems.

Rouse, W. B. (1986b). Man-machine systems. In J. A. White (Ed.), *Production handbook* (Chapter 23). New York: Wiley.

Rouse, W. B. (1990). *The dynamics of mental workload.* Norcross, GA: Search Technology, Inc.

Rouse, W. B., and Cody, W. J. (1986). Function allocation in manned system design. *Proceedings of the 1986 IEEE International Conference on Systems, Man, and Cybernetics,* pp. 1600–1606.

Rouse, W. B., Geddes, N. D., and Curry, R. E. (1987). An architecture for intelligent interfaces: Outline of an approach to supporting operators of complex systems. *Human-Computer Interaction, 3,* 87–122.

Rouse, W. B., and Johnson, W. B. (1989). *Computational approaches for analyzing tradeoffs between training and aiding.* Brooks Air Force Base, TX: Air Force Human Resources Laboratory.

Sewell, D. R., Rouse, W. B., and Johnson, W. B. (1989). *Initial evaluation of principles for graphical displays in maintenance problem solving.* Arlington, VA: Office of Naval Research.

Sheridan, T. B. (1984). Supervisory control of remote manipulators, vehicles, and dynamic processes: Experiments in command and display aiding. In W. B. Rouse (Ed.), *Advances in man-machine systems research* (Vol. 1, pp. 49–137). Greenwich, CT: JAI Press.

Sheridan, T. B., and Ferrell, W. R. (1974). *Man-machine systems: Information, control, and decision models of human performance.* Cambridge, MA: MIT Press.

Walden, R. S., and Rouse, W. B. (1978). A queueing model of pilot decision making in a multi-task flight management situation. *IEEE Transactions on Systems, Man, and Cybernetics, SMC-8,* 867–875.

Wickens, C. D. (1984). Processing resources in attention. In R. Parasuraman and R. Davies, (Eds.), *Varieties of attention.* New York: Academic.

Wickens, C. D., Tsang, P., and Pierce, B. (1985). The dynamics of resource allocation. In W.B. Rouse, (Ed.), *Advances in man-machine systems research* (Vol. 2, pp. 1–49). Greenwich, CT: JAI Press.

Wierwille, W. W., Casali, J. G., Connor, S. A., and Rahimi, M. (1985). Evaluation of the sensitivity and intrusion of mental workload estimation techniques. In W. B. Rouse (Ed.), *Advances in man-machine systems research* (Vol. 2, pp. 51–127). Greenwich, CT: JAI Press.

Chapter **6**

The Sales and Service Phase

Initiation of the sales and service phase signals the accomplishment of several important objectives. The product or system will have been successfully tested, verified, demonstrated, and evaluated. In addition, the issues of viability, acceptability, and validity will have been framed, measurements planned, and initial measurements executed. These initial measurements, beyond the framing and planning, will have exerted a strong influence on the nature of the product or system.

The above efforts will have been accomplished using the approaches, methods, and guidelines in Figure 6.1. This list is not meant to be comprehensive. The criterion for choosing items to appear on this list was simply that a summarizing figure appears in Chapters 1 through 5. Also, many additional approaches, methods, and guidelines are summarized in figures in later chapters.

It is useful to reflect briefly on the nature of the items listed in Figure 6.1. This set of structured issues, phases, methods, steps, and so forth, is intended to provide a clear, but *nominal,* path for pursuing human-centered design. The word "nominal" is underscored to emphasize that this sequence does not have to be followed in lockstep in order to assure success. For any particular design effort, some aspects of this sequence may not be relevant, or more appropriate context-specific approaches may be available. Nevertheless, in general, I have found that these approaches, methods, and guidelines provide good starting points on the path to design success.

- Seven Measurement Issues (Fig. 2.1)
- Four Phases of Measurement (Fig. 2.8)
- Four Aspects of Organizing for Measurement (Fig. 2.9)
- Eight Methods and Tools for Measurement (Figs. 3.1 and 4.2)
- Four Buying Influences (Fig. 4.1)
- Twelve Guidelines for User Acceptance (Figs. 4.6a–c)
- Four Steps of Design/Documentation (Fig. 5.2)
- Three Levels of Evolutionary Architectures (Fig. 5.3)
- Three Steps/Activities of Function Allocation (Figs. 5.4 and 5.5)
- Six Analogies of Human Behavior and Performance (Fig. 5.7)

Figure 6.1. Summary of approaches, methods, and guidelines.

SALES AND SERVICE ISSUES

In this phase, one is in a position to gain closure on viability, acceptability, and validity. One can make the measurements necessary to determining if the product or system really solves the problem that motivates the design effort, solves it in an acceptable way, and provides benefits that are greater than the costs of acquisition and use. This is accomplished using the measurement plan that was framed in the naturalist phase, developed in the marketing phase, and refined in the engineering phase.

These measurements should be performed even if the product or system is "presold"—for example, when a design effort is the consequence of a winning proposal. In this case, even though the "purchase" is assured, one should pursue closure on viability, acceptability, and validity in order to gain future projects.

There are several other activities in this phase beyond measurement. One should assure that the implementation conditions for the product or system are consistent with the assumed conditions upon which the design is based. This is also the point at which the later steps of user acceptance plans are executed, typically with a broader set of people than those who participated in the early steps of the plan. This phase also involves some of the later aspects of technology transition plans that are discussed in Chapter 10.

The sales and service phase is also where problems are identified and remediated. To the greatest extent possible, designers should work with users and customers to understand the nature of problems and alternative solutions. Some problems may provide new opportunities rather than indicate shortcomings of the current product or system. It is important to recognize when problems go beyond the scope of the original design effort. The emphasis then becomes one of identifying mechanisms for defining and initiating new design efforts to address these problems.

The sales and service phase also provides an excellent means for maintaining relationships. One can identify changing buying influences that occur because of promotions, retirements, resignations, and reorganizations. Moreover, one can lay the groundwork and make initial progress on the naturalist phase, and perhaps the marketing phase, for the next project, product, or system.

METHODS AND TOOLS FOR MEASUREMENT

How does one make the final assessments of viability, acceptability, and validity? Furthermore, how does one recognize new opportunities? Unstructured direct observation can provide important information. However, more formal methods are likely to yield more definitive results and insights. Figure 6.2 lists the methods and tools appropriate for answering these types of questions.

METHODS AND TOOLS	PURPOSE	ADVANTAGES	DISADVANTAGES
Sales Reports	Assess perceived viability of product or system.	The ultimate, bottom-line measure of success.	Information on lack of sales due to problems is likely to be too late to help.
Service Reports	Assess typical problems and impact of solutions attempted.	Associate problems with customers and users and enable follow-up.	May be too late for major problems and may not explain cause.
Questionnaires	Query large number of customers and users regarding experiences with product.	Large population can be inexpensively queried.	Low return rates and shallow nature of responses.
Interviews	In-depth querying of small number of customers and users regarding experiences with product.	Face-to-face contact allows in-depth and candid interchange.	Difficulty of gaining access, as well as time required to schedule and conduct.

Figure 6.2. Methods and tools for measurement.

Sales Reports

Sales are an excellent measure of success and a good indicator of high viability, acceptability, and validity. However, at the risk of beating a dead horse, I have to emphasize again that sales reports are a poor way of discovering a major design inadequacy. In addition, when a major problem is detected in this manner, it is quite likely that one may not know what the problem is or why it occurred.

Service Reports

Service reports can be designed, and service personnel trained, to provide much more than simply a record of service activities. Additional information of interest concerns the specific nature of problems, their probable causes, and how users and customers perceive and react to the problems. Users' and customers' suggestions for how to avoid or solve the problems can also be invaluable. Individuals' names, addresses, and telephone numbers can also be recorded so that they subsequently can be contacted.

Questionnaires

Questionnaires can be quite useful for discovering problems that are not sufficient to prompt service calls. They also can be of use for uncovering problems with the service itself. If a record is maintained of all customers and users, this population can regularly be sampled and queried regarding problems, as well as ideas for solutions, product upgrades, and so on. As noted before, though, a primary disadvantage of questionnaires is the typical low return rate.

Interviews

Interviews can be a rich source of information. Users and customers can be queried in depth regarding their experiences with the product or system, what they would like to see changed, and new products and systems they would like. This can also be an opportunity to learn how their organizations make purchase decisions, both in terms of decision criteria and budget cycles.

While sales representatives and service personnel can potentially perform interviews, there is great value in having designers venture out to the sites where their products and systems are used. However, these ventures should not be unstructured "tours." There should be clear measurement goals, questions to be answered, an interview protocol, and so on, much in the way that is described in Chapters 3 and 4.

Summary

The sales and service phase brings the measurement process full circle. An important aspect of this phase is using the above tools and methods to initiate the next iteration of naturalist and marketing phases. To this end, as was emphasized in Chapter 3, a primary prerequisite at this point is the ability to *listen.*

LEVERAGE POINTS

The remainder of this book is concerned with four topics: evaluation, aiding, training, and technology transition. Each of the topics is treated at length—a full chapter each—because of the potential high returns for investing in these areas. The "returns" of interest concern the extent to which a product or system is human-centered.

Because of the potential payoff in these areas, I call them leverage points. Figure 6.3 summarizes the ways in which each of these points provides leverage. This summary is elaborated in the remainder of this chapter.

POINT	APPROACH TO LEVERAGE
Evaluation	Utilization of a sequence of evaluation methods that efficiently and effectively provides answers to evaluative questions in a timely manner and enables satisfaction of system requirements within budget and schedule constraints.
Aiding	Human limitations can be overcome and their abilities enhanced by providing additional system functionality specifically targeted and designed to go beyond baseline functionality in a manner that is predominantly focused on human-centered objectives.
Training	Early consideration of knowledge and skills necessary to successful use of a product or system can lead to design modifications, as well as training concepts, that substantially enhance perceptions of the product or system and potentially decrease life-cycle costs.
Technology Transition	Successful adoption and use of a product or system are often linked to understanding and resolving organizational and other people-related issues that can expedite the transition of R&D results to design, and design concepts to the marketplace.

Figure 6.3. Leverage points.

Evaluation

Chapter 7 discusses evaluation. The efficiency and effectiveness of evaluation are crucial. If evaluation is too expensive, too slow, and/or too late, it is quite likely that it will be done poorly or, in effect, skipped completely. More to the point, evaluation provides the basis for assuring that a product or system will satisfy requirements. Shortchanging this aspect of measurement portends consequences that are anything but good.

Aiding

The design of aiding is considered in Chapter 8. Aiding is included in a design to make tasks easier, enable levels of performance that would not otherwise be achievable, or in some cases facilitate performance when it would not otherwise be possible. In general, aiding is added to a baseline design in order to assure that the human–machine system can meet performance requirements without unacceptable workload or other undesirable side effects. This orientation makes aiding a very important means, a leverage point, for developing a human-centered design.

Training

The design of training is discussed in Chapter 9. In contrast to aiding, which provides a means of directly augmenting human performance, training is a process of providing people with the potential to perform. Of special concern early in the process of designing a product or system are the knowledge and skills that will have to be taught in order to have personnel able to perform as needed.

Knowledge and skill requirements dictate training requirements in terms of programs of instruction, training devices, and the overall training system. Important considerations here are the cost of training, the time that training requires, and the availability of people whose aptitudes, abilities, and attitudes are such that they can be trained to requisite levels of performance. To the extent that cost and/or time are unacceptable, or people are unavailable, either the system has to be redesigned or aiding provided so that less training is needed. Trade-offs such as this, which should be performed early, provide an important leverage point for assuring a human-centered design that meets requirements within budget and schedule constraints.

Technology Transition

Technology transition is the topic of Chapter 10, the final chapter in this book. In earlier discussions, it was noted that user acceptance was a necessary condition for successfully transitioning a product or system into use. From the broader perspective of Chapter 10, user acceptance remains necessary, but is no longer sufficient for success.

Many types of individuals and organizations are stakeholders in the process that produces a new product or system and puts it in use. It is essential that one understand and resolve a variety of people-related issues concerned with transitioning R&D results to design applications, and design inventions to marketplace innovations. The understanding of these issues and the ability to resolve them are the ultimate leverage points for successful human-centered design.

Chapter 7

System Evaluation

Evaluation is the process of determining whether or not a product or system meets requirements. If a system does not meet requirements, an evaluation effort should also provide means for determining why requirements were not met. Thus, evaluation should not be viewed as simply a pass/fail measurement process.

As emphasized in Chapters 2 and 5, it is useful to view evaluation in terms of three issues (Rouse, 1984, 1986a; Rouse et al., 1984):

- *Compatibility* is the extent to which the nature of physical presentations to the user and the responses expected from the user are compatible with human input–output abilities and limitations.
- *Understandability* is the extent to which the structure, format, and content of the user–system dialogue result in meaningful communication.
- *Effectiveness* is the extent to which the system leads to improved performance, makes a difficult task easier, or enables accomplishing a task that could not otherwise be accomplished.

Ideally, evaluation should be a process of first resolving compatibility issues, then understandability issues, and finally effectiveness issues. What is needed are efficient and useful methods for achieving this ideal.

MEASUREMENT METHODS

Figure 7.1 illustrates how multiple methods can be sequenced to pursue the three evaluation issues. This figure is not meant to imply that all four types of evaluation are required for every evaluation effort. For example, if a full-scope simulator is not available, one could substitute an enhanced part-task simulator and/or increase the extent of in-use evaluation.

The sequential relationships among the methods shown in Figure 7.1, as well as a variety of feedback or iteration loops not shown, were chosen to provide design-oriented successive refinement in an efficient manner. Therefore, the methods that are employed first are those that are relatively

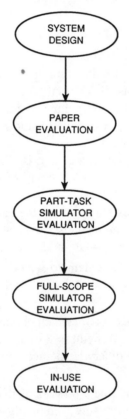

Figure 7.1. Evaluation with multiple methods.

fast and inexpensive, and that can be employed earliest in the design process. Consequently, one would avoid, for example, an early transition directly from system design to in-use evaluation. Such a transition would risk investing a large amount of resources to learn that compatibility and/or understandability deficiencies preclude assessment of effectiveness. While this caution may seem so obvious as to be unnecessary, I observed one full-scope simulator evaluation where over one million dollars was invested to learn that operators had difficulty reading the displays—a *very* expensive way to uncover a compatibility problem.

Paper Evaluation

The primary purpose of a paper evaluation is to assess compatibility in the sense of determining the degree to which a system design takes advantage of people's abilities while avoiding their limitations. The prerequisites or inputs required for a paper evaluation include a working prototype, available design documentation on objectives and requirements, and access to designers to obtain information that is not documented. Scale drawings, rather than a prototype, are sufficient for a partial, formative paper evaluation, but not for a complete formal paper evaluation. This is because drawings, or even static pictures of displays, are not sufficient for assessing the full range of compatibility issues.

The results of a typical paper evaluation will include a list of problems identified and recommendations for modifying the design to eliminate the problems and assure the compatibility of the system. The range of compatibility issues is quite large. Compilations can be found in a variety of textbooks (e.g., Kantowitz and Sorkin, 1983), handbooks (e.g., Rouse, 1986b), and industrial standards (e.g., NUREG-0700 [Nuclear Regulatory Commission, 1981]).

In order to expand the scope of a paper evaluation to include understandability and perhaps a few effectiveness considerations, one must carefully analyze the design to see that it satisfies information requirements and to assess, if possible, the degree to which system objectives are achieved. If a design process such as outlined in Chapter 5 has been used, a paper evaluation of understandability is unlikely to yield many surprises and often can be avoided. In the absence of a systematic design process, however, such a paper evaluation may be quite valuable, particularly if it forces the formulation of information requirements—"backing in" to information requirements almost inevitably leads to design modifications.

We developed a methodology for performing a paper evaluation of understandability (Rouse et al., 1984). It involves analyzing the information displays, input devices, and dialogue characteristics to determine the knowledge requirements for meaningful communication. An assessment is then made of the extent to which potential users have this knowledge. Knowledge deficits are addressed by redesign to eliminate these requirements, incorporation of information on the displays to satisfy the requirements, and/or training to provide the requisite knowledge.

A formal paper evaluation of compatibility and/or understandability should not begin until *after* a fully tested and verified prototype is available. Only then can evaluation of a design concept be a fully meaningful exercise. Of course, this principle does not preclude qualitative analyses and informal evaluations of tentative concepts. Indeed, such activities are essential to integrating design and evaluation. However, one should carefully avoid using formal evaluations as a means for catching incomplete or inadequate engineering and/or programming.

Once a paper evaluation of compatibility has been completed and the results are used as a basis for modifying the design, the focus of evaluation shifts to understandability and effectiveness. As this transition occurs, one should be in a position to feel confident that compatibility is no longer an issue.

Part-Task Simulator Evaluation

It is quite common for evaluations to be performed using full-scope simulators, especially if such a simulator is readily available. This approach certainly has a high degree of face validity. But this approach can be very expensive, particularly if the primary concerns are compatibility and understandability. These issues should be resolved before one reaches the costly point of full-scope simulator evaluation, or even in-use evaluation.

On the other hand, a paper evaluation is clearly insufficient for resolving all evaluation issues. Somewhere in the evaluation process, people have to use the system, and the results of this usage must be determined. Considering the expense associated with using a full-scope simulator for evaluation, it seems reasonable to compromise, at least initially, on face validity. A part-task simulator embodies such a compromise.

A part-task simulator is a device that roughly approximates the real system of interest in terms of appearance, responses, and inputs. A wide range of part-task simulators is possible, bracketed by static mock-ups on the low end and full-scope simulators on the high end. The extent to which a part-task simulator reflects the real system—which is called simulator fidelity—can be low if part-task evaluation will eventually transition to full-scope evaluation. However, if full-scope evaluation is not feasible, one may need a higher-fidelity part-task simulator so that a wider range of issues can be pursued with this device.

For designs where the dynamic response of the system is of central importance, there are several ways of implementing dynamic response capabilities in a part-task simulator. One approach is to compute responses using a mathematical model of the dynamic process. This approach is often used for full-scope simulators. To be feasible for part-task simulators, though, various approximations are typically employed to lessen computational complexity. As might be expected, these approximations inevitably result in a decrease in dynamic fidelity relative to the actual process of interest. Fortunately, modest losses of fidelity should not be problematic, especially when evaluating understandability.

Another common approach to part-task simulation is to tape-record the responses of a full-scope simulator and then replay them on the part-task simulator. This approach has the obvious advantage of maintaining dynamic fidelity while decreasing computational requirements. It also can be relatively inexpensive.

On the other hand, this approach has the very significant disadvantage that people using the part-task simulator cannot influence the response that they are viewing. This disadvantage can be partially overcome by having people take "hypothetical" actions. However, it is very improbable that this mechanism will be sufficient for people to feel *totally* involved in the process. For this reason, part-task simulators based on tape playback are mainly useful for assessing understandability. To assess effectiveness, people have to be able to influence the course of events.

Regardless of the type of part-task simulator employed, the primary objective of part-task simulator evaluation is to assess understandability. This involves determining whether people can comprehend the information transmitted to them by the system and whether they can communicate their desires, and perhaps their intentions, to the system. These determinations involve several questions:

- What information will the system typically transmit to users?
- What will users need to know in order to comprehend this information?

- What information will users typically transmit to the system?
- What will users need to know in order to formulate and communicate this information?
- How will the knowledge requirements noted above be satisfied by training, documentation, and/or the system itself?

These questions should have been asked and answered during design rather than evaluation. The purpose of evaluating understandability is to assess the validity of the answers generated during design by answering the following questions:

- Do users actually comprehend the information presented by the system?
- Do users correctly formulate inputs to the system?
- Do users correctly communicate inputs to the system?

In general, the results of a part-task simulator evaluation should include a thorough assessment of understandability and perhaps an initial evaluation of effectiveness. In addition to an assessment of understandability, results should include identification of the specific deficiencies that must be eliminated to achieve an acceptable level of understandability. Many deficiencies will be resolvable by design changes.

It is not unusual, however, for deficiencies to provide insights into inadequacies of users' training. It may be necessary to augment their training for the new system. It is very useful to know this early—in this case, in the engineering phase—so that training requirements can be anticipated rather than discovered once the system is in use.

Once paper evaluations and part-task simulator evaluations are complete, and the system and/or training plan has been modified accordingly, the focus of evaluation shifts to effectiveness. It is important as one transitions to full-scope simulator or in-use evaluation that compatibility and understanding no longer be issues.

Full-Scope Simulator Evaluation

The primary purpose of full-scope simulator evaluation is rigorous confirmation that requirements have been satisfied—that the system is effective. If one has used paper evaluations and part-task simulator evaluations to assess and resolve compatibility and understandability issues, one should not have the problem of suddenly discovering during full-scope simulator evaluation that the displays are unreadable, the messages incomprehensible, the training inadequate, or performance measures ill-defined. In contrast, if one chooses to move directly from design to full-scope simulator evaluation without any intermediate phases of evaluation, the risk of problems such as these is substantial. Beyond wasting resources, a premature full-scope simulator evaluation that results in these problems can lead to a whole design concept being discredited, even if problems are only due to a lack of compatibility. From this perspective, incompatibilities are far more than simply "little details."

As noted above, full-scope simulator evaluations should focus on effectiveness (e.g., the extent to which decision making is supported) rather than understandability (i.e., the interpretability and expressibility of messages), which is best pursued in a part-task simulator. Because the use of full-scope simulators can be quite expensive, one should attempt to limit both the number of users involved and the length of time each user spends in the simulator. This can be accomplished in two ways.

First, most training can be accomplished using the part-task simulator. There is no reason to use a full-scope simulator to train users to remember color-coding schemes or use the dialogue structure of a system. The practice needed to become skilled in the use of such information can easily be pursued in a part-task simulator. In fact, there is some evidence that a full-scope simulator may "get in the way" of learning new concepts—this issue is discussed in Chapter 9.

A second way to ration the use of the full-scope simulator is to limit the range of conditions studied. This can be achieved by exploring the full range of interesting issues in a part-task simulator and using the results obtained to determine one or two crucial conditions needing study in the full-scope simulator. As noted in Chapter 5, models can also be used to explore and filter possible variations of conditions.

Since the emphasis of both full-scope simulator and in-use evaluations is effectiveness, it is essential that operationally useful measures of effectiveness be defined *prior* to measurement. A common practice is to collect every data element imaginable and worry about measures later. This not only wastes resources, but also frequently leads to the unfortunate discovery after the evaluation that not every data element was imagined! This problem can be lessened if, when defining requirements during design, one also considers how to measure the degree to which requirements are satisfied.

Examples of operationally useful measures of effectiveness include errors, inefficiency, and the amount of time required, for instance, to detect, diagnose, and compensate for problems. These and other performance measures are discussed in considerable detail in later sections of this chapter. The particular choice of measures can become a point of negotiation among the stakeholders in a design effort. Guidelines for making such choices are presented later in this chapter.

In-Use Evaluation

Full-scope simulator evaluation is usually the final step of formal evaluation. In this case, in-use evaluation is basically a follow-up effort aimed at observing use of the system and gathering information for the purpose of design modifications and extensions. The nature of this type of in-use evaluation was discussed in Chapter 6, which dealt with the sales and service phase.

In situations where a full-scope simulator is not available or justifiable, formal in-use evaluations can be performed. While such an evaluation can have a high degree of face validity, rather sizable evaluation problems can arise. It may be difficult to create the conditions that are most important to evaluate—these conditions may, for example, be too dangerous to create. Data acquisition may also be difficult and limited to what is operationally available rather than what is experimentally desirable.

Summary

Figure 7.2 summarizes much of the above discussion of measurement methods in terms of the usefulness and efficiency of alternative methods. Usefulness concerns the extent to which a method is appropriate for addressing compatibility, understandability, or effectiveness. Efficiency refers to the time and other resources necessary to address an evaluation issue using a particular method.

Recommended methods for each evaluation issue are in italic. Recommendations are based on usefulness and efficiency. Note that effectiveness evaluations are inherently inefficient if one employs methods that are fully useful.

Figure 7.2, as well as the related material presented earlier in this section, can provide the basis for developing an evaluation plan. Detailed guidelines for planning evaluations can be found in Rouse (1984) and a variety of textbooks on design of experiments (e.g., Montgomery, 1984).

METHOD OF MEASUREMENT	EVALUATION ISSUE		
	COMPATIBILITY	UNDER-STANDABILITY	EFFECTIVENESS
Static Paper Evaluation	*Useful and Efficient*	Somewhat Useful but Inefficient	Not Useful
Dynamic Paper Evaluation	*Useful and Efficient*	Somewhat Useful but Inefficient	Not Useful
Data-Driven Part-Task Simulation	Useful but Inefficient	*Useful and Efficient*	Marginally Useful but Inefficient
Model-Driven Part-Task Simulation	Useful but Inefficient	*Useful and Efficient*	*Somewhat Useful but Efficient*
Full-Scope Simulation	Useful but Very Inefficient	Useful but Inefficient	*Useful but Somewhat Inefficient*
In-Use Evaluation	Useful but Extremely Inefficient	Useful but Very Inefficient	*Useful but Inefficient*

Figure 7.2. Usefulness and efficiency of alternative measurement methods.

To conclude this section, it is important to note that some of the concepts discussed in this section (e.g., simulators) have to be interpreted differently depending on the context of the design effort. For aviation, process control, and power production, the idea of a simulator is common. In contrast, for design domains such as office automation, for example, the notion of a word processor simulator may sound unusual.

However, the concept is meaningful for word processors. One might develop a version of a word processor that is "instrumented" to capture keystroke and timing data, as well as control the nature of the typing and editing tasks. If the scope of this simulator included only a portion of the task of work processing, perhaps only those aspects that were novel and needed in-depth evaluation, the device would be a part-task simulator.

CHOOSING METHODS AND MEASURES

The previous section focused on methods for evaluating compatibility, understandability, and effectiveness. The next logical issue is the choice of measures. This section considers the interaction between methods and measures.

To understand this interaction, almost 200 past studies of complex human–machine systems were reviewed (Rouse and Rouse, 1984). For the purposes of this analysis, each study or experiment reviewed represented a single piece of data. To be included in the sample, a study had to involve a real system, full-scope simulator, or reasonably high-fidelity part-task simulator. Studies involving abstract laboratory tasks or games were not included in the sample.

Of the nearly 200 studies initially identified, 115 were selected for further analysis. The other studies were eliminated for two reasons. The emphasis of some studies was purely methodological and thus, the descriptions were intentionally incomplete. For many studies, especially those reported in conference papers and technical reports, the reporting of methods, measures, and so on, was inadequate for classifying the study.

Classification of Studies

The 115 studies were classified using the scheme shown in Figure 7.3. The categories in the figure that are entitled domain of study, method, and approach to measurement are self-explanatory. The remaining categories deserve further discussion.

With regard to type of data, virtually every study measured system outputs (i.e., what people saw, heard, etc.) and system inputs (i.e., what people did). A few studies recorded verbal conversations or protocols of "thinking aloud." Several evaluations involved subjective assessments of the system state or the overall situation, either by the system users themselves or by observers. Preferences of users were frequently assessed. Occasionally, a priori characteristics of users (e.g., level of experience) were determined.

In terms of data collection methods, most studies employed computer (or equivalent) logs or transcripts of outputs and inputs. Audiotape, videotape, observers' notes, and users' notes were occasionally used to supplement the logs or transcripts. Questionnaires were fairly common and interviews rare.

- Domain of Study
    Vehicle Control
    Process Control
    System Maintenance
    Other
- Method
    Actual Equipment
    Full-Scope Simulator
    Part-Task Simulator
    Other
- Approach to Measurement
    Controlled Experiment
    Field Study of Performance
    Survey of Opinion
- Type of Data
    System Outputs/Inputs
    Verbal Communications/Protocols
    Subjective Assessments/Preferences
    Subject Characteristics
- Data Collection Methods
    On-line Objective Measurement
    On-line Subjective Assessment
    Off-line Measurement/Assessment
- Measures Assessed
    Cumulated and/or Averaged
    Sampled
    Transformed
    Disaggregated
    Subjective
- Nature of Results
    Planned Comparison—Effect Significant
    Planned Comparison—Effect Not Significant
    Evaluation of Approach
    Description

Figure 7.3. Overall classification scheme.

For measures assessed, most studies involved measures that were cumulated or averaged over time or across trials. Measures in this cumulated/averaged category included such factors as time, errors, cost, trials, observations, actions, inquiries, variability. In general, each of these measures resulted in a single aggregate number being assigned for an entire trial.

A few evaluations measured performance at only a single instant for an entire trial. In some cases, the instant of interest was defined as the time when the maximum or minimum value of a specific variable occurred throughout the course of a trial. In other cases, the instant of interest reflected a particularly crucial point in a trial (e.g., touchdown for an aircraft).

Several studies transformed the data measured into an aggregate index of performance. Examples included cumulated weighted squared errors, information gain, input–output transfer functions, and input–output stability metrics. These types of measure resulted in physical variables being mapped to more abstract metrics.

A variety of studies considered measures where aggregation was avoided. Typically, errors, costs, actions, and so on, were disaggregated into categories and/or time periods. Multiple analyses were then performed in an attempt to gain insight into the process whereby aggregate performance resulted.

Subjective measures were fairly common. These measures included users' opinions and preferences as well as their assessments of their own workload. Often, these types of measure were used for corroboration rather than as primary measures.

The nature of results category in Figure 7.3 indicates that the majority of the studies involved planned comparisons of displays, aiding schemes, training methods, or task characteristics. Most of these planned comparisons resulted in statistically significant differences being demonstrated. The remainder of the planned comparisons provided insufficient evidence to conclude that a difference existed. Those studies not involving planned comparisons typically resulted in an evaluation of an approach to experimentation, or simply a description of users' behaviors and preferences.

Analysis of Studies

The goal of the classification effort described above was to determine the extent to which specific combinations of methods, measures, and so forth, tend to lead to definitive evaluative results. Results were termed definitive if a significant difference was demonstrated. Results were termed nondefinitive when the investigators were forced to accept the "null hypothesis" that no difference existed.

It is important to emphasize that acceptance of the null hypothesis is usually a weak conclusion for a study. All one can rigorously claim in such a situation is that observed differences were insufficient in magnitude to justify concluding that these differences were due to other than chance. One would not want to choose methods and measures that tend to lead to accepting the null hypothesis.

To pursue this issue, the studies involving planned comparisons were cross-tabulated relative to the subcategories of each of the other major categories in Figure 7.3. The proportion of studies accepting the null hypothesis was then determined for each of the subcategories. Pairwise statistical comparisons of proportions were performed using appropriate statistics.

Resulting Guidelines

Before discussing the resulting guidelines, it first must be noted that these results are correlative in the sense that they do not reflect the results of a controlled assessment of the effects of methods and measures. Consequently, it could be that all of the results reflect the fact that "good" evaluators tend to use particular methods and measures, and "bad" evaluators tend to use others. While such a phenomenon is possible, it is not probable. Further, given that practical guidance must be based on whatever

---

1. Full-scope and part-task simulator methods are more likely to yield definitive results, particularly for studies in the domains of vehicle control and other.
2. Disaggregate measures are more likely to yield definitive results, particularly for studies in the domains of vehicle control and other.
3. If aggregate measures must be used, simulator methods are preferred, with part-task simulators being less problematic than full-scope simulators.

Figure 7.4. Guidelines for choosing methods and measures.

One is more likely to obtain definitive experimental results if the method chosen allows a reasonable degree of control and the measures chosen allow fine-grained analyses of performance.

Figure 7.5 General principle for choosing methods and measures.

information is available, it is possible to summarize the results of these analyses to yield a few guidelines for the range of methods and measures studied.

The resulting three guidelines are shown in Figure 7.4. In general, these recommendations are most important for the domains of vehicle control and others, such as manipulators, management, and command and control. However, considering the fact that the size of the sample of studies within the domains of process control and maintenance was much smaller than that for vehicle control, it may be prudent to follow the guidelines in Figure 7.4 for all domains.

By considering the nature of the three guidelines, it is possible to suggest a general principle that is shown in Figure 7.5. To a great extent, this principle is intuitively obvious. But based on the foregoing analysis and results, one can argue for this principle much more forcefully than if its only basis were intuition. This can be especially important when one is requesting resources for evaluation and this request is competing with requests of other stakeholders in the design process.

DIMENSIONS OF HUMAN PERFORMANCE

It was noted earlier in this chapter that a common practice in complex evaluative efforts is to collect sufficient data for every measure imaginable and subsequently decide which measures are really important. This practice carries two risks. One is the risk of being overwhelmed by the resulting data. The other is the possibility that delaying the decision of what is important may result in necessary data not being collected.

Obviously, the best strategy is to collect only the data that is important. The difficulty, of course, is choosing among the myriad of possibilities. However, it is quite likely that many of the possibilities are not independent—they are just different ways of measuring the same thing.

To pursue this possibility, we analyzed data collected for 60 aircraft power plant mechanics who performed troubleshooting tasks using two maintenance training simulators (Henneman and Rouse, 1984). A total of 30 measures were considered. Twelve of these related to a priori characteristics of mechanics in terms of standard tests of ability, aptitude, and cognitive style. The remaining 18 measures related to troubleshooting performance. This set of 18 included every measure that we could find that had been utilized in previously published studies of troubleshooting.

The resulting data for 60 mechanics and 30 measures were analyzed using a variety of statistical techniques. In considering the measures of a priori characteristics of mechanics, it was found that cognitive style was moderately correlated with performance and, when combined with measures of ability, the correlation with performance was fairly high. With regard to the 18 measures of performance, it was found that there were only three unique dimensions of performance: time, errors, and inefficiency.

Time is a very straightforward metric. Errors are more subtle than might be imagined—this dimension is discussed in detail in the next section. The concept of inefficiency provides a way of characterizing behaviors that are not erroneous and do not necessarily require excessive time, but nevertheless are not the best choices among behaviors. For example, in troubleshooting, inefficient actions are those that are productive in the sense that these actions move the diagnostic process in the direction of isolating the failure, but are inefficient because other choices would have been more productive.

The idea that there are only three dimensions of human performance is very powerful. While it cannot be asserted that the results of the analysis of aircraft mechanics troubleshooting performance are valid for other jobs and tasks, this is probably a reasonable conclusion from a purely practical point of view. Accordingly, for any evaluative effort, particularly for effectiveness, one should make sure that the measures chosen cover these three dimensions.

It is quite likely, of course, that measures of system performance will also be needed. These measures tend to be very context-specific and usually reflect the "bottom line" for evaluation. From this perspective, it might be argued that one only need assess the bottom-line measures. However, in light of the guidelines discussed in the last section, it is prudent to also collect fine-grained measures of the three dimensions of human performance. Such fine-grained measures can be valuable when diagnosing the source of less-than-acceptable levels of the bottom-line measures.

Of the three dimensions of human performance, errors are perhaps the richest source of insights. This is due to the rather straightforward and somewhat aggregate nature of the time dimension, as well as the highly context-specific nature of the inefficiency dimension. Errors, in contrast, can be disaggregated in very interesting ways, and understanding of the nature of errors has relevance across many contexts.

ANALYSIS AND CLASSIFICATION OF HUMAN ERROR

Human error is often cited as the probable cause of accidents or un-acceptable performance in systems such as aircraft, ships, and power plants. For many of these systems, it has been estimated that 70 to 90 percent of all accidents can be traced to human error. Furthermore, as discussed above, error is a unique and important dimension of human performance. Hence, the problem of human error is both substantive and important.

Two major approaches can be taken to characterizing human error: probabilistic and causal. The probabilistic approach is typically pursued by those who are interested in the human reliability aspects of risk analysis. In these analyses, human error is treated in a manner quite similar to that used for hardware failures. Failure rates for humans for particular types of task, as well as failure rates for hardware components, serve as inputs to an analysis that produces an overall reliability metric. The use of this approach is often dictated by requirements that system reliability meet some specified level.

In contrast, the causal approach to characterizing human error is based on the premise that errors are seldom random and, in fact, can be traced to causes and contributing factors that, once isolated, can perhaps be eliminated or at least ameliorated. Thus, the causal approach can be useful for evaluating and subsequently modifying system designs and, as is later illustrated, training programs. Succinctly, the causal approach is concerned with compensating for errors rather than counting them, which is the primary concern of the probabilistic approach.

This section presents a methodology for analysis and classification of human error that is premised on the causal approach (Rouse and Rouse, 1983). The purpose of this methodology is to provide a practical design and evaluation tool. The practical value of this methodology is illustrated by discussions of four applications.

Classification Scheme

Numerous error classification schemes have been proposed. The scheme discussed in this section builds on many of these previous efforts, particularly the work of Rasmussen (1986).

The internal consistency and generality of a classification scheme are likely to be enhanced if the scheme is based on a model of the process within which errors occur. Such a model can help to identify categories within the classification scheme and also illustrate the relationships among categories. Figure 7.6 represents a simplified view of the tasks humans perform in systems such as aircraft, ships, power plants, and manufacturing facilities.

During normal operations, the human is assumed to cycle through observing the system state, choosing procedures, and executing procedures. Procedures need not be formal in the sense of checklists. For example, unwritten "scripts" that unconsciously govern much routine behavior can be viewed as procedures.

When the system state is such that one or more state variables have values outside the normal range, or when warnings or alarms have been activated, the situation is deemed abnormal and the human may have to resort to some form of problem solving. If the abnormality is one that the human has frequently experienced, patterns may be recognized, the solution may be immediately obvious, and the human may go directly to choosing a procedure. On the other hand, if the pattern of state variables, warnings, and alarms is unfamiliar to the human, a mode of problem solving that relies less on pattern recognition may be necessary.

The choosing and testing of hypotheses will usually result in at least tentative identification of the source of the problem. As identification proceeds, the human may be faced with choosing among alternative goals or perhaps coordinating conflicting goals. For instance, in operational situations (as opposed to maintenance) the goal of continued operational availability may conflict with the goal of avoiding compromises in safety or losses of equipment. Once a goal is chosen, the human must choose a procedure for achieving the goal and then execute the procedure.

The relationships among tasks depicted in Figure 7.6 are not meant to be strictly interpreted. For example, the flow of control among tasks is certainly more complex than this figure represents. Nonetheless, this conceptual model provides a means for organizing an error classification scheme.

The resulting human error classification scheme includes two levels: general categories and specific categories. General categories discriminate among the behavioral processes within which human errors occur. Specific categories define the particular characteristics of erroneous decisions or actions.

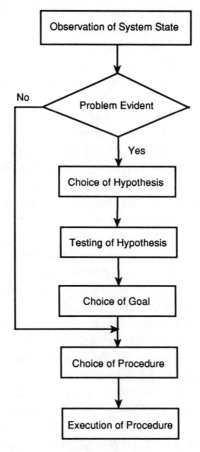

Figure 7.6. Conceptual model of humans' tasks.

The general categories of the classification scheme correspond to the processes shown in Figure 7.6. For any given domain of study, some of these general categories may not be relevant. For example, in a maintenance task "choice of goal" may not be needed because the diagnostic goal would not be competing with the operational goal. As another illustration, studies of human error during normal operations would not need to consider the general categories of "choice of hypothesis" and "testing of hypothesis" because the human would not be involved in problem solving.

Therefore, the general classification scheme shown in Figure 7.7 has to be specialized for particular task domains. Alternatively, one can utilize all categories with the expectation that errors in some categories will inherently not occur. However, as is later discussed, this approach can present problems when testing the statistical significance of differences in error frequencies.

Many of the specific categories in Figure 7.7 are self-explanatory. For instance, the general terms incorrect, incomplete, inappropriate, unnecessary, and lack are quite straightforward. Brief definitions of the less general terms appear in Figure 7.8. The terms excessive and misinterpreted require additional explanation.

Excessive observation of system state refers to repeated checking to see that the values of the state variables are the same as observed earlier. In domains such as process control, some checking and cross-checking are reasonable and desirable. However, in other domains it is an unproductive use of time and may indicate a lack of understanding or confidence on the part of the human. From this perspective, one can see that the quantitative definition of excessive depends on the task domain.

Misinterpretation of observation of system state refers to erroneous interpretation of correct readings of the right set of state variables. An example is reading a variable correctly but interpreting its value as acceptable when in fact it is too high. Another example is reading a variable correctly but using it as an indicator of the state of an unrelated subsystem.

The classification scheme shown in Figure 7.7 should be used as follows. First, as noted earlier, the general categories relevant to the task domain of interest are determined using Figure 7.6. Then, within each general category, the relevant specific categories are determined. These determinations depend on the nature of the task domain, the purpose of the study, and the types of data available. As a result, there was considerable variation in the general and specific categories employed in the four applications discussed later in this chapter.

GENERAL CATEGORY	SPECIFIC CATEGORY
1. Observation of System State	a. excessive b. misinterpreted c. incorrect d. incomplete e. inappropriate f. lack
2. Choice of Hypothesis	a. inconsistent with observations b. consistent but very unlikely c. consistent but very costly d. functionally irrelevant

3. Testing of Hypothesis	a. incomplete
	b. false acceptance of wrong
	   hypothesis
	c. false rejection of correct
	   hypothesis
	d. lack

4. Choice of Goal	a. incomplete b. incorrect c. unnecessary d. lack
5. Choice of Procedure	a. incomplete b. incorrect c. unnecessary d. lack
6. Execution of Procedure	a. step omitted b. step repeated c. step added d. steps out of sequence e. inappropriate timing f. incorrect discrete position g. incorrect continuous range h. incomplete i. unrelated inappropriate action

Figure 7.7. Human error classification scheme.

SPECIFIC CATEGORY	BRIEF DESCRIPTION
1a. excessive	improper rechecking of correct readings of appropriate state variables
1b. misinterpreted	erroneous interpretation of correct readings of appropriate state variables
1c. incorrect	incorrect readings of appropriate state variables
1d. incomplete	failure to observe sufficient number of appropriate state variables
1e. inappropriate	observations of inappropriate state variables
1f. lack	failure to observe any state variables
2a. inconsistent	could not cause particular values of state variables observed
2b. unlikely	could cause values observed but much more likely causes should be considered first
2c. costly	could cause values observed but very costly (in time or money) place to start
2d. irrelevant	does not functionally relate to state variables observed
3a. incomplete	stopped before reaching a conclusion
3b. acceptance	reached wrong conclusion
3c. rejection	considered and discarded correct conclusion
3d. lack	hypothesis not tested
4a. incomplete	insufficient specification of goal
4b. incorrect	choice of counterproductive goal
4c. unnecessary	choice of nonproductive goal
4d. lack	goal not chosen
5a. incomplete	choice would not fully achieve goal
5b. incorrect	choice would achieve incorrect goal
5c. unnecessary	choice unnecessary for achieving goal
5d. lack	procedure not chosen
6a. omitted	required step omitted
6b. repeated	unnecessary repetition of required step
6c. added	unnecessary step added
6d. sequence	required steps executed in wrong order
6e. timing	step executed too early or too late
6f. discrete	discrete control in wrong position
6g. continuous	continuous control in unacceptable range
6h. incomplete	stopped before procedure complete
6i. unrelated	unrelated inappropriate step executed

Figure 7.8. Definitions of specific human error categories.

Data Collection

To study human error in depth, one usually needs a substantial amount of information about the process leading up to an error. This information includes time histories of displayed variables and the human's actions. To help in interpreting these objective data, more subjective data such as observers' notes, transcripts of verbal communications, and verbal protocols are quite useful and perhaps necessary. Interviews and questionnaires can also be helpful.

In general, if one hopes to identify the probable cause of errors and contributing factors, one should try to collect data from multiple objective and subjective sources in order to provide corroborating evidence. While the collection of this variety of data can create logistical difficulties, such technical problems are minor compared to those associated with data analysis, a topic that is discussed later.

Another important consideration in studies of human error is the choice of human subjects to study. If one is interested in the design and evaluation of systems such as aircraft, ships, power plants, and factories, then one should study the errors of humans who actually operate these types of systems. In some situations, it may be possible to isolate one or more subtasks and study naive users performing these subtasks, with appropriate training. This is a common practice in research laboratories.

However, two important risks are associated with studying human error in this way. First, in order to be able to employ naive users, one may end up studying a subset of system operations or a subtask that is insignificant relative to overall human performance. In other words, one risks the possibility of abstracting to the point where the real task is lost. A second potential difficulty is associated with the fact that naive users may have quite different aspirations, motivations, and attitudes toward risks than professional users. For example, if college students are employed as naive users, they may try to avoid all errors, while professional users may accept a few errors with the knowledge that this allows them to concentrate more on optimizing system operations, as well as with the knowledge that they will quickly detect and recover from most errors. Thus, a variety of considerations dictates that one be quite careful when deciding to study naive users.

Identification of Errors

The identification of errors involves analyzing the various types of data discussed above and, for a first pass, locating the errors of omission and commission. This requires that one define all of the decisions that should have been made and actions that should have been taken and vice versa. These definitions provide a filter for locating potential errors. One must then decide whether or not the omission or commission is actually an error, just a difference in preferences, or possibly irrelevant.

This initial filtering, if one accepts that it is only a first pass, is not too difficult. Typically, multiple experts are used to screen the raw data. Somewhat algorithmic approaches to error identification have been employed for well-defined tasks. For constrained task domains, where procedures or scripts prescribe much of behavior, it is possible to automate the initial filtering (e.g., Hammer, 1984).

The results of the filtering process must then be analyzed to identify errors. For highly structured tasks, this can perhaps be done algorithmically—a method for accomplishing this is discussed in Chapter 8. If an algorithmic approach is precluded, multiple independent experts should be used to decide if each omission and commission is actually an error. In addition, if one is performing a comparison of system designs or training programs, the independent experts should be "blind" or unaware of the particular system design or type of training associated with a specific anomaly that is to be judged as erroneous or not.

Quite likely, not all experts will agree about each error, although the nature of human–machine interaction is such that widespread disagreement is unlikely. Discussion of disagreements often leads to the conclusion that one or more of the experts has a technical misunderstanding that, once clarified, will lead to resolution of the disagreement. For the remaining disagreements, which usually are few in number, majority-rule voting or some other decision rule can be used as a basis to proceed.

Identification of Causes and Contributing Factors

If one is concerned with differences in system designs or training programs, then planned comparisons are likely to be possible. One can employ methods of experimental design (Montgomery, 1984; Rouse, 1984) and assess the degree to which one design or program produces significantly more errors than the other. If one finds significant differences, one typically concludes that the experimental manipulations (e.g., the differences in designs or programs) caused the differences in number of errors. This conclusion may lead to subsequent experiments where these causes of errors are explored in more detail.

Many studies of human error do not involve planned comparisons. Instead, some are concerned with evaluating a specific system design or training program. In situations such as these, one can attempt to find correlates of human error that lead to hypotheses regarding causes. These correlates can be viewed as factors that may contribute to the occurrence of human errors.

Since the nature of potential contributing factors tends to be highly context-dependent, specifying a general classification of contributing factors is difficult. However, some guidelines can be offered for defining potential contributing factors. Four very general classes of factors are

- Inherent human limitations,
- Inherent system limitations,
- Contributing conditions, and
- Contributing events.

Inherent human limitations include the knowledge, skills, and attitudes of the people of interest—see, for example, Kantowitz and Sorkin (1983) for general discussions of human limitations. Knowledge and skills may be based on either training or experience and encompass, for example, knowledge of basic system operations as well as more advanced knowledge such as the functioning of automatic systems. The need to discriminate among types of knowledge and skill depends on the purpose of the study. Attitudinal factors may include such things as complacency, automation "mind-set," and interpersonal relationships among users. Knowledge, skills, and attitudes may be assessed via analysis of interviews, questionnaires, transcripts of communications, and verbal protocols. Multiple sources of this type of information are often useful. Analysis of linguistic information may require the use of formal methods such as content analysis or linguistic analysis.

Inherent system limitations include the design of controls and displays, design of dialogues and procedures, and level of simulator fidelity in those cases where simulators are employed. Determining if basic human engineering principles have been violated is fairly straightforward—basically it involves performing compatibility and understandability analyses of the task environment within which the error data were collected. Determining whether or not any inadequacies identified could have potentially affected a particular observation, decision, and action sequence is also fairly easy. However, as with the other contributing factors, determining a particular inadequacy to be *the* cause of an error is by no means easy.

Contributing conditions may include environmental factors such as noise; excessive workload; frustration, anger, or embarrassment; confusion; and operating in degraded modes. Conditions such as noise and degraded-mode operations are reasonably easy to assess. Methods of assessing workload were discussed in Chapter 5. Subjective assessments of confusion, frustration, and so on, can be provided during the course of data collection by trained observers, or can be provided after the fact by the people involved. Alternatively, or in addition, experts can analyze interviews, transcripts of communications, and verbal protocols.

Contributing events may involve distractions, lack of communication, misleading communication, sudden equipment failures, and more subtle events such as tension release (e.g., laughter or swearing). Some of these events can be objectively identified, whereas others require analysis of linguistic information.

Assessing the degree to which many of these contributing factors are present is often a very subjective matter. Hence, the use of multiple independent experts is very important. The assumption is, of course, that agreement among experts denotes truth. While this may be a tenuous assumption in general, it seems to be reasonable in domains such as aviation, shipping, power production, process control, and manufacturing.

Classification of Errors

After agreement has been reached with regard to the omissions and commissions that will be termed errors, the next step is error classification. While one could rely on the original data for making classification decisions, this can be extremely tedious and time-consuming, especially since each expert evaluator will have already poured through all of the data in the process of identifying errors.

To avoid this difficulty, it is convenient to develop a one-paragraph description of each error that includes all information relevant to the error from all data sources (i.e., observation and action histories, transcripts of communications, etc.). These descriptions should be such that all of the experts agree that the set of information is complete. Once the set of descriptions is finished, each expert is furnished with a package including, for each error, a description, an error classification form, and a contributing factor assessment form. Each expert then independently classifies and assesses contributing factors for all errors.

As with error identification, the classification of errors can result in disagreement among experts. These disagreements can be resolved in the same ways as used for disagreements in identification.

Statistical Analysis

Analysis of planned comparisons of system designs or training programs can be performed using traditional statistical tests. The relationships between contributing factors and frequencies of errors can be assessed using correlation methods. Measures of relationship can also be used to determine the extent of agreement or concordance among independent expert evaluators.

It is important to note that statistical procedures can add a degree of rigor to a topic that otherwise can be overly subjective. However, it is equally important to emphasize that the statistics are not ends in themselves. They should mainly serve as means to generating new ideas in the process of evaluating and improving the design of systems and training programs.

Remediation of Errors

As emphasized at the beginning of this chapter, the purpose of identifying errors is not simply to count them. The goal is to remediate the problems underlying the occurrence of errors. This can be done in two ways. One alternative is to eliminate errors by removing their cause. The second alternative is to allow errors to happen but compensate for their consequences—this approach is called error tolerance and is discussed in Chapter 8.

Either of these two approaches is enhanced by having a deeper explanation of the nature of the errors. Two types of explanation are needed. One type is *task-oriented*. This type of explanation is reflected in the error classification scheme in Figures 7.7 and 7.8, as well as the discussions of contributing factors.

The other type of explanation is *psychologically oriented* and provides a behavioral diagnosis of an error. Norman (1981) and Reason and Mycielska (1982) have devised an especially useful behavioral dichotomy. It enables one to distinguish between humans' intentions and their execution of actions.

If the intention is appropriate for the situation, but the execution is incorrect, the error is termed a *slip*. Types of slip include attentional capture, misperceptions, and losing track of one's place (Reason and Mycielska, 1982).

In contrast to slips, if the intention is inappropriate, even though the execution may be correct with respect to this intention, the error is termed a *mistake*. Reason (1983) has suggested that types of mistakes include oversimplifications (bounded rationality), appearances of uncalled schema (imperfect rationality), overreliance on familiar cues and well-worn solutions (reluctant rationality), and irrational acts. In other words, mistakes can result from human information processing limitations, processing errors, or unwillingness to invest the effort necessary to formulate intentions more carefully.

The distinction between slips and mistakes has practical implications for how errors are remediated. Errors such as unintentional closing of the wrong valve or pushing the wrong button are likely to be immediately corrected if the system is designed to provide the feedback necessary for humans to detect these events. On the other hand, doggedly correct execution of an inappropriate emergency recovery strategy may be difficult to remediate so simply. In the following case studies, the nature of slips and mistakes is further elaborated. Chapter 8 continues the discussion in the context of error tolerant interfaces.

CASE STUDIES

The four case studies discussed in this section occurred over a five-year period. During this period, the methodology described in the previous section continued to evolve. Various aspects of the methodology were extended and refined as we learned from each study. In this section, no attempt is made to resurrect earlier versions of the methodology. These earlier versions are discussed in the referenced publications cited in the descriptions of each case study.

Operation of Supertanker Propulsion Systems

This study involved seven crews of professional engineering officers who were being trained using a high-fidelity supertanker engine control room simulator (van Eekhout and Rouse, 1981). Troubleshooting exercises were observed. Measurement methods included verbal protocols, computer logs of all discrete events, interviews, questionnaires, and observer ratings.

The resulting data were analyzed for human errors by two independent analysts. Errors were independently classified using an early version of the classification scheme in Figures 7.7 and 7.8. Contributing factors were also assessed, including lack of knowledge of system functioning and automatic controller functioning, human factors design inadequacies, and simulator fidelity inadequacies.

It was found that errors associated with inappropriate identification of failures (i.e., false acceptance of wrong hypothesis) were highly correlated with a lack of knowledge of the functioning of the basic system as well as the automatic controllers within the system. These errors tended to be mistakes (rather than slips) in that inappropriate intentions were formed based on inadequate or incorrect knowledge.

Errors related to execution of procedures were highly correlated with inadequacies of the layout of the control panel and simulator fidelity inadequacies. These errors tended to be slips due to, for example, control knobs that were easily confused or labels that were difficult to read.

The types of slips that occurred were anticipated, but the nature of the mistakes provided a new insight. Errors in diagnosing failures of automatic controllers were often associated with a lack of understanding of failure modes of the controllers. In these situations, the cues presented by the automation were misleading. Crew members became confused and/or reached incorrect conclusions upon which they proceeded to act. Succinctly, basic misunderstandings led to mistakes. This phenomenon was encountered throughout this series of studies. Such findings provide a clear basis for modification of training programs.

Maintenance of Aircraft Power Plants

This effort included two studies involving 58 trainees in a Federal Aviation Administration certificate program in aircraft power plant maintenance (Johnson and Rouse, 1982). These studies involved comparing three methods for training troubleshooting. Two of the methods included computer-based power plant simulations. The third used instructional television and a fairly traditional lecture and demonstration format.

The data collected in these studies were analyzed in a similar manner to the supertanker study. An important difference, however, was the error classification scheme. The aircraft maintenance tasks of interest were sufficiently different from supertanker operations to require a shift of emphasis in the use of the classification scheme in Figures 7.7 and 7.8. In particular, the categories eliminated related to choice of goal and those emphasized related to providing a finer-grained view of troubleshooting errors. This tailoring of the classification scheme to the domain of interest was found to be very important in order to avoid having many categories where the error frequencies are inherently zero. For some types of statistical analysis, a large number of zero frequencies will mask significant differences among nonzero frequencies.

The results of analyzing the data from the first of the two experiments showed traditional video-based instruction to be superior. For the most part, this was demonstrated by procedural errors by the mechanics who were trained with the computer-based methods. Specifically, it was found that mechanics trained with the low- to moderate-fidelity computer simulations knew *what* troubleshooting tests to make, but not *how* to make them.

This conclusion led to combining the two computer-based methods and adding material on how to make tests. The second study in this effort compared this hybrid computer-based method to the traditional instruction. Results indicated that the previous types of errors no longer occurred, and the two methods of instruction yielded similar performance.

This effort showed how fine-grained analysis of performance can lead to insights and improvements that might not be evident from global performance measures. In particular, lack of understanding of test procedures led to execution errors. These errors are probably best categorized as mistakes. This insight helped to modify the training to provide the requisite understanding.

Operation of Aircraft Flight Information System

This study involved four two-person crews flying a high-fidelity, twin-engine aircraft simulator (Rouse, Rouse, and Hammer, 1982). Each crew flew several full-mission, commuter airline scenarios. The purpose of the study was to evaluate an onboard computer-based system for retrieving, displaying, and assisting in executing aircraft procedures. This system was the first of a series of concepts that culminated in the intelligent cockpit discussed in earlier chapters.

The data collected were analyzed using methods similar to those for the previous studies. The components of the error classification scheme (Figures 7.7 and 7.8) associated with procedure execution were emphasized.

The results of the error analysis indicated that errors of omission (i.e., leaving out procedural steps) were virtually eliminated. In this manner, the interface was, to an extent, slip-tolerant. However, errors of commission (i.e., adding unnecessary steps) were relatively unaffected by the computer-based information system. The reason was simple—the computer had no way of discriminating incorrect actions from irrelevant actions.

A subsequent reanalysis of this data (Rouse and Rouse, 1983) indicated that errors of omission were often associated with possible confusion, distraction, and/or communication problems. In contrast, errors of commission did not have any obvious causes and contributing factors. It is important to understand the basis for extra actions that are not necessarily incorrect. This topic is revisited during the discussion of error tolerant interfaces in Chapter 8.

Operation and Maintenance of Diesel Generators

This study involved evaluating the impact of computer-based training for operation and maintenance of auxiliary diesel generators in nuclear power plants (Maddox, Johnson, and Frey, 1986). This study is discussed in detail in Chapter 9. Accordingly, only the results of the error analysis are noted here.

It was found that it was not possible to discriminate computer-based training from traditional instruction using only overall performance measures—this finding is quite consistent with the guidelines shown in Figure 7.5 and the principle summarized in Figure 7.6. However, an error analysis led to the conclusion that traditional instruction resulted in 500 percent more consequential errors than computer-based instruction. This substantial difference was found to persist even 20 weeks after training had been completed. Without an error analysis, this important impact of computer-based training would not have been identified.

SUMMARY

Evaluation is the process of determining whether or not the system design satisfies requirements. Early and iterative evaluation performed efficiently is an important leverage point in human-centered design.

Evaluation often results in identifying shortfalls where requirements are not satisfied—hopefully, these shortfalls are discovered during early evaluations rather than late ones. Many of these problems can be resolved by design modifications. Not infrequently, though, what is needed is one or both of the leverage points that are next discussed—aiding and training.

REFERENCES

Hammer, J. M. (1984). An intelligent flight management aid for procedure execution. *IEEE Transactions on Systems, Man, and Cybernetics, SMC-14,* 885–888.

Henneman, R. L., and Rouse, W. B. (1984). Measures of human problem solving performance in fault diagnosis tasks. *IEEE Transactions on Systems, Man, and Cybernetics, SMC-14,* 99–112.

Johnson, W. B., and Rouse, W. B. (1982). Analysis and classification of human errors in troubleshooting live aircraft power plants. *IEEE Transactions on Systems, Man, and Cybernetics, SMC-12,* 389–393.

Kantowitz, B. H., and Sorkin, R. D. (1983). *Human factors: Understanding people-system relationships.* New York: Wiley.

Maddox, M. E., Johnson, W. B., and Frey, P. R. (1986). *Diagnostic training for nuclear plant personnel.* Volume 2: *Implementation and evaluation* (Rept. NP-3829, Vol. 2). Palo Alto, CA: Electric Power Research Institute.

Montgomery, D. C. (1984). *Design and analysis of experiments.* New York: Wiley.

Norman, D. A. (1981). Categorization of action slips. *Psychological Review, 88,* 1–15.

Nuclear Regulatory Commission (1981). *Guidelines for control room design reviews* (NUREG-0700). Washington, DC: U.S. Nuclear Regulatory Commission.

Rasmussen, J. (1986) *Information processing and human-machine interaction: An approach to cognitive engineering.* New York: North-Holland.

Reason, J. (1983). On the nature of mistakes. In N. Moray and J. W. Senders (Eds.), *Reprints of NATO Conference on Human Error.*

Reason, J., and Mycielska, K. (1982). *Absent minded: The psychology of mental lapses and everyday errors.* Englewood Cliffs, NJ: Prentice-Hall.

Rouse, S. H., Rouse, W. B., and Hammer, J. M. (1982). Design and evaluation of an onboard computer-based information system for aircraft. *IEEE Transactions on Systems, Man, and Cybernetics, SMC-12,* 451–463.

Rouse, W. B. (1984). *Computer-generated display system guidelines.* Volume 2: *Developing an evaluation plan* (Rept. NP-3701, Vol. 2). Palo Alto, CA: Electric Power Research Institute.

Rouse, W. B. (1986a). Design and evaluation of computer-based decision support systems. In S. J. Andriole (Ed.), *Microcomputer decision support systems* (Chapter 11). Wellesley, MA: QED Information Systems.

Rouse, W. B. (1986b). Man-machine systems. In J. A. White (Ed.), *Production handbook* (Chap. 23). New York: Wiley.

Rouse, W. B., Kisner, R. A., Frey, P. R., and Rouse, S. H. (1984). *A method for analytical evaluation of computer-based decision aids* (Rept. NUREG/CR-3655). Oak Ridge, TN: Oak Ridge National Laboratory.

Rouse, W. B., and Rouse, S. H. (1983). Analysis and classification of human error. *IEEE Transactions on Systems, Man, and Cybernetics, SMC-13*, 539–549.

Rouse, W. B., and Rouse, S. H. (1984). A note on evaluation of complex man-machine systems. *IEEE Transactions on Systems, Man, and Cybernetics, SMC-14*, 633–636.

van Eekhout, J. M., and Rouse, W. B. (1981). Human errors in detection, diagnosis, and compensation for failures in the engine control room of a supertanker. *IEEE Transactions on Systems, Man, and Cybernetics, SMC-11*, 813–816.

Chapter **8**

Design of Aiding

A wide range of definitions of aiding are available. In this book, aiding is defined as functionality that is added to a baseline design concept specifically for the purpose of enhancing human decision making, problem solving, and performance in general. This definition does not encompass the baseline functionality associated with using a product or system. Thus, for example, an automatic transmission in an automobile represents aiding; the steering wheel does not. As another illustration, remote control for a television represents aiding; the screen and on/off button do not.

This definition of aiding allows for the possibility of the baseline evolving in time. For instance, the baseline automobile of today is much different than the baseline of 50 years ago. In this way, typically over a long period of time, aspects of aiding transition to the baseline and are expected to be part of the product or system.

Traditionally, at least for 40 years or so, aiding was referred to as decision aiding. This supposedly limited aiding to supporting making choices or judgments. Roughly 20 years ago, it was recognized that choices and judgments occur in a broader context. From this realization, the notion of decision support systems emerged.

In recent years, the concept of decision support has become very broad. Decision support is now viewed as providing assistance for

- Formulation, analysis, and interpretation (Sage, 1981),
- Information retrieval and management, analysis and reasoning, representation, and judgment (Zachary, 1986), and
- Situation assessment, planning, commitment, execution, and monitoring (Rouse, 1986).

From this broader view, the concepts of decision aiding and decision support might best be thought of as task or job aiding and support. This perspective underlies this chapter in that aiding is conceptualized in terms of supporting tasks and jobs rather than only choices and judgments.

A fundamental dichotomy exists among many proponents of aiding and support systems. One side of this dichotomy, the *prescriptive* side, views aiding as a means of assuring that humans perform as they should. The other side, the *descriptive* side, views aiding as a means of enabling people to do what they want to do and enhancing their performance. The descriptive approach is feasible with well-trained and motivated people. In contrast, the prescriptive approach may be necessary with poorly trained and less motivated people. Clearly, training and aiding interact. This interaction and the resulting trade-offs are discussed in Chapter 9.

Another distinction within the realm of aiding and support systems, concerns the basis from which human behavior and performance are conceptualized. Hammond, McClelland, and Mumpower (1980) cluster the many paradigms and points of view on this issue into two groups.

One group has its roots in economics and views humans as behaving so as to maximize subjective expected gain. In terms of the notion of constrained optimality discussed in Chapter 5, this group models humans as gain maximizers subject to constrained abilities to estimate and update probabilities, as well as related human limitations.

The other group has its roots in psychology and focuses on the features and patterns people perceive and how this information, perhaps unconsciously, influences their behaviors and performance. This group models humans as "satisfiers" in the sense that humans are assumed to seek satisfactory, but not necessarily optimal, courses of action.

In this chapter, neither point of view is embraced. Probabilities, expected values, features, and patterns do not explicitly play a role in the concepts and methods discussed in sections subsequent to the next section, where traditional approaches to aiding are discussed. Instead, these later sections focus squarely on the design of aiding to enhance task and job performance.

Therefore, this chapter does not take a position on two of the major dichotomies within the community of researchers and designers of aiding and support systems. However, it is quite likely that stakeholders in any particular effort to develop an aiding or support system will have strong opinions on the issues underlying these dichotomies. It is important in the naturalist and marketing phases to identify and understand these opinions. Otherwise one could potentially design a system based on tacit assumptions about human behavior and performance that are in conflict with the users and/or customers.

TRADITIONAL APPROACHES TO AIDING

Before discussing several concepts for comprehensively aiding and supporting tasks and jobs, it is useful to review a representative set of more traditional approaches to aiding. This brief review provides a basis for understanding a more comprehensive taxonomy of alternative aiding concepts that is presented in a later section of this chapter where a methodology for designing aiding and support systems is outlined and illustrated.

Problem Formulation/Structuring

For some types of problems, especially those where a reasonable amount of information is available and time constraints are not very tight, humans' performance can be enhanced by assisting them in formulating or structuring problems. This assistance often takes the form of means for explicating alternatives, defining attributes of consequences, and assessing alternatives in terms of these attributes.

This information is often structured by using a *decision tree* that includes choice points, where the human has alternatives to choose among, and consequence points where the laws of probability govern the outcomes. Leal and Pearl (1977) have developed an interactive computer program that assists people in "growing" decision trees.

An alternative representation of this type of information is a *plan goal graph*. Such a representation enables decomposition of top-level goals into lower-level subgoals, plans for accomplishing goals, and actions that compose plans. Pearl, Leal, and Saleh (1982) have developed an interactive computer program for supporting this form of goal-directed decision structuring.

Assistance in formulating and structuring problems can often be quite valuable. It is not unusual for the best decision to be obvious once a problem is structured, mitigating any need to pursue subsequent evaluation and selection. The primary limitation of this type of assistance is the frequent occurrence of situations where not all of the viable alternatives are known, information on consequences is unavailable, and time constraints are relatively tight.

Probability Estimation and Updating

A wide variety of psychological studies in the 1950s and 1960s showed that people's abilities to estimate probabilities are limited. In addition, people have difficulty accurately updating probabilities in light of new information or evidence. These studies, and many more in the 1970s, provided many insights into humans' biases and limitations as processors of probabilistic information.

A natural conclusion to reach from these results is that people need assistance in estimating and updating probabilities. A variety of aiding schemes emerged. The aid developed by Adelman, Donnell, Phelps, and Patterson (1982) is a good example. To use this computer program, the user defines the alternative hypotheses and estimates the a priori probability associated with each alternative. Then, as new information becomes available, the user estimates the probability that this information would result if each hypothesis were true. The computer then uses appropriate rules of probability to update the probability that each hypothesis is true in light of the information received thus far.

Aids that function in this manner usually result in much more accurate probability estimates and updates. The primary limitation of this approach is that there are very few tasks and jobs where people have to explicitly use probabilities. In those few cases where they do—for example, weather forecasting—people have been found to be reasonably proficient at estimating and updating. Hence, this type of aiding is of most value in situations where people have to use probabilities, but have little experience in this type of activity.

Selection Among Alternatives

If the relevant alternatives are known, their attributes determined, attribute levels specified, perhaps probabilistically, and a multiattribute criterion function defined, then it is usually feasible to program a computer to determine the best or optimal alternative. Computers can do this much more consistently and much faster than humans.

Aiding of this type has been the most prevalent form of aiding. The work of Freedy and his colleagues, which is summarized in Freedy, Madni, and Samet (1985), is a good representative of this approach. Their method involves assessing individual decision makers' multiattribute criterion functions by inferring the function that best fits observations of the individual's choices. A computer then uses this function to make choices for the individual. Several applications of this method have shown that more consistent decisions and faster decisions are made using this method of assistance.

The primary limitation of this approach is the requirement that the decision-making situation be highly structured and a considerable amount of information be available. There are several types of important decision where these prerequisites are satisfied. However, in general, the degree of structure and quantification necessary are not feasible.

Execution and Monitoring

The above types of aiding provide assistance in defining alternatives and attributes, estimating and updating probabilities associated with alternatives, and choosing among alternatives. In many situations, alternatives represent courses of action. In these cases, the next step after choosing an alternative course of action is executing the sequence of actions.

Execution and monitoring aids assist in performing actions and monitoring the consequences. A good example of such an aid is the system discussed in Chapter 7 that assists in retrieving, executing, and monitoring aircraft procedures (Rouse, Rouse, and Hammer, 1982). To the extent that onboard hardware and software allow, this system will execute procedural steps for the crew and/or monitor the crew's execution of steps. The monitoring component of this aid is a special case of a broader concept called an *error-tolerant interface* (Rouse and Morris, 1987). This concept is discussed in more detail later in this chapter.

The feasibility of execution and monitoring aids depends on having courses of action that are well defined, as well as the availability of appropriate sensors and actuators so that the aid can "observe" and "act." Within systems such as aircraft, ships, power and process plants, and factories, these prerequisites are often satisfied. In less engineering-oriented systems such as offices, shops, hospitals, and so on, feasibility is often less likely, particularly due to the prohibitive costs of the necessary technology for assisting execution and monitoring in loosely structured environments.

ADAPTIVE AIDING

The value of aiding is apt to be judged relative to users' unaided performance and, in particular, users' perceptions of their unaided performance. Further, the perceived value of aiding involves a trade-off between potential performance improvements and the effort involved in learning and utilizing aiding. Perceived ease of use can be an enormously significant factor.

Perceptions of performance and ease of use can be strongly affected by the multitask nature of realistically complex systems. Unaided human performance in multitask situations depends on the mix of task demands at any point in time. As a result, aiding for any given task may sometimes be of value and at other times unnecessary. At those points in time when aiding is most needed, it is likely that users will have few resources to devote to interacting with the aiding. Put simply, if a user had the resources to instruct and monitor the aiding, he or she would probably be able to perform the task with little or no aiding.

The above perspective easily leads to the conclusion that the level of aiding, as well as the ways in which users and the aiding interact, should change as task demands vary. More specifically, the level of aiding should increase as task demands become such that human performance will unacceptably degrade without aiding. Moreover, the ways in which users and the aiding interact should become increasingly steamlined as task demands increase. Finally, it is quite probable that variations in levels of aiding and modes of interaction will have to be initiated by the aiding rather than by users whose excess task demands have created a situation requiring aiding. The term *adaptive aiding* is used to denote aiding concepts that meet the requirements outlined in this paragraph.

Evolution of the Concept

The concept of adaptive aiding emerged in the mid-1970s in the course of a project that was concerned with applications of artificial intelligence (AI) to cockpit automation. In the process of conducting several pilot-in-the-loop demonstrations of AI modules for onboard diagnosis of equipment failures, the possibility of "conflicting intelligence" was discovered. For example, in the course of diagnosing and compensating for a particular system failure, pilot and computer could produce mutually counterproductive actions.

The need for "cooperative intelligence" became patently obvious. Although it is typical for humans to shoulder the burden of such cooperation, this possibility was inconsistent with the design philosophy of the project. The computer, therefore, would have to "understand" the pilot sufficiently to be able to ensure cooperation. In other words, the computer's expertise would have to include understanding of human behavior and performance within the domain of interest.

This goal motivated an R&D program that continues today, more than 15 years after its initiation (Rouse, 1988b). As the adaptive aiding concept, and associated technology, have matured, this approach to aiding became integral to the intelligent cockpit discussed in Chapters 3 and 4, and described in more detail later in this chapter. Five studies by different individuals and organizations have clearly demonstrated the performance benefits of adaptive aiding.

These five studies cover only a portion of the range of possibilities for adaptive aiding outlined in the next section entitled Framework for Design. In general, these studies involved model-based adaptation (i.e., aid-initiated adaptation based on predictions of human performance models) to particular users for the purpose (in four of the studies) of allocating tasks to either human or computer. The wider range of possibilities for adaptive aiding is discussed in the aforementioned section on design.

An early study involved a multitask flight management scenario with adaptive task allocation (Chu and Rouse, 1979). During roughly the same period, Freedy and his colleagues studied adaptive aiding in an antisubmarine warfare (ASW) intelligence-gathering task (Freedy, Madni, and Samet, 1985). Revesman studied adaptive task allocation in a process control task (Revesman and Greenstein, 1986). Morris studied adaptive task allocation in an aerial reconnaissance task (Morris, Rouse, and Ward, 1988), and Forester (1986) studied adaptive task transformation (i.e., variable display formats) for the same task.

In all of these studies, adaptive aiding was shown to yield performance improvements relative to either manual performance or nonadaptive aiding. Chu found improvements ranging from 2 to 40 percent in terms of reduced response time to flight management task demands. Freedy and his coworkers found that decisions regarding ASW sensor placements were made 15 percent faster. Revesman's results indicated a 73 percent decrease in the number of redundant actions. Morris's studies of aerial reconnaissance resulted in 5 to 9 percent improvements in manual control performance and 22 to 25 percent improvements in target identification. For this same reconnaissance task, Forester found that variable formats yielded 12 to 25 percent improvements in manual control and 18 to 42 percent improvements for target identification.

All five of these applications involved adaptation on the basis of measured or predicted human performance. Aiding was provided, via recommendation or preemption, when human performance was predicted to be unacceptable without aiding. Two central elements of this type of adaptation are human performance models and on-line assessment methods.

Underlying Models and Methods

In order to be able to adapt the nature of aiding to optimize (or at least improve) performance, it is necessary to predict how well a human is likely to perform a particular task at a specific point in time. Although in principle an empirical data base could be developed to provide this information, this approach is totally impractical for any realistically complex application. Accordingly, models are needed whereby on-line predictions of performance can be obtained based on the current state of task demands and the availability of human sensorimotor and information processing resources. These models represent one of the ways in which expertise about human behavior and performance can be embedded in an intelligent support system.

A variety of human performance models have emerged in the course of developing and evaluating adaptive aiding concepts. Task allocation was originally conceptualized as a multiserver queueing problem involving two time-shared computers between which tasks were allocated in a dynamic manner. By adding appropriate behavioral constraints, one of the two "computers" in this formulation represented human multitask performance (Walden and Rouse, 1978). This model provided the basis for determining the parameters of Chu's adaptive allocation scheme (Chu and Rouse, 1979).

Although a queueing model is useful for predicting which tasks will be performed and how much time will be required, such models do not inherently predict how well tasks will be performed. Greenstein developed a model utilizing pattern recognition methods and queueing theory that, at least partially, provides this capability for process monitoring tasks

(Greenstein and Rouse, 1982; Greenstein and Revesman, 1986). Govinda-raj developed a model based on control theory that was quite successful in predicting timing and performance effects of coordinating continuous and discrete tasks (Govindaraj and Rouse, 1981).

The efforts by Greenstein, and especially Govindaraj's work, clearly demonstrated that on-line performance modeling can present enormous computational problems. As a result, the concept of a "matrix of models" emerged whereby a variety of simple models are indexed and accessed in terms of tasks and performance metrics (Rouse, 1981; Rouse, Geddes, and Curry, 1987). An example of the type of model that might be within the matrix is Morris's regression model of target recognition performance (Morris, Rouse, and Ward, 1988). It is of particular note that this model is highly context-specific, computationally trivial, and very accurate for the aerial reconnaissance task of concern.

This model is also interesting because it was found that predictions of task performance should not only be based on impending task demands but also reflect recently completed task demands. This can result in a "carry-over" effect whereby aiding may be suitable for a task that is usually easy because the human has just completed performance of a difficult task. The implication of this result is that a static or time-invariant view of human performance modeling may be inadequate for many adaptive aiding ap-plications.

A variety of types of human performance models are potentially of use within adaptive aiding. The issue of primary concern is not one of choos-ing or developing the best or most appropriate model. Instead, the concern is with judiciously integrating aspects of various models. From this per-spective, some of the expertise embedded within an adaptive aiding system involves modeling expertise about ranges of applicability and interpreta-tion of human performance models.

Beyond predicting human performance and anticipating degradations, adaptive aiding requires on-line assessments of what the human is doing and, if possible, what the human intends to do (Rouse, 1977). These requirements reflect the multitask nature of the systems of interest. Were the concern with single-task situations, the activities and intentions of humans would be rather obvious. However, multitask situations lead to potential conflicts in terms of task competition between human and com-puter, as well as tasks "dropping through the cracks."

Several early efforts represented humans' intentions in terms of meeting task demands (e.g., Enstrom and Rouse, 1977; Greenstein and Rouse, 1982; Govindaraj and Rouse, 1981). It was assumed that humans would not have goals other than those imposed by immediate task demands. This is clearly an inappropriate assumption in many operational contexts.

Geddes (1985, 1989) has addressed this issue for fighter aircraft pilots and, to a lesser extent, for process control operators. His approach involves interpreting humans' actions—switch flips, button pushes, and control movements—in terms of a "natural" language of aircraft piloting or process controlling. This enables interpretation of these actions in the context of scripts, plans, and goals (Schank and Abelson, 1977) appropriate for the domain of interest.

As might be expected, the accuracy and utility of the resulting interpretations are directly dependent on the degree of structure in the task environment. When the degree of discretion allowed humans precludes inference of intentions, it is always possible to communicate explicitly with the humans involved if time allows. Fortunately, in all of the situations thus far investigated, time-response constraints and degree of task structure appear to be highly correlated—when explicit communication is needed, it is typically the case that time for such communication is available.

Another approach to on-line assessment is what might be termed "leading indicators" methods. For example, in Morris's aerial reconnaissance experiments, it was found that the average time to detect a target, once it came into the field of view, tended to start increasing about 20 seconds prior to any targets being missed. Thus, detection time or latency, which was not a measure of particular importance, served as a leading indicator of the primary measure of detection accuracy. While this particular measure is obviously specific to the task studied, the notion of identifying leading indicators is very attractive.

Framework for Design

The conceptual design of an adaptive aid can be approached systematically by pursuing answers to a specific set of design questions or issues. While it is not possible to provide generic, context-free answers to these questions, it is possible to outline the range of alternative answers and provide principles of adaptation and interaction to assist designers in choosing among these alternatives (Rouse, 1988b). These questions and alternative answers are summarized in Figure 8.1 and elaborated in the remainder of this section.

What Is Adapted To? In general, adaptation to the user and/or task is possible. In addition, an aid can adapt to a class of users (or tasks), a particular user (or task), or a particular user (or task) at a specific point in time. In other words, adaptation can be relative to a class as a whole, a member of a class, or the state of a particular member.

- What is Adapted To?
  - Class of Users or Tasks
  - Member of Class
  - Member at Specific Point in Time
  - Adapting *to* vs. Adapting *of*

- Who Does the Adapting?
  - Designer
  - User
  - System

- When Does Adaptation Occur?
  - Off-line
  - On-Line in Anticipation of Changes
  - On-line in Response to Changes

- What Methods of Adaptation Apply?
  - Transformation (making a task easier)
  - Partitioning (performing part of a task)
  - Allocation (performing all of a task)

- How is Adaptation Done?
  - Measurement
  - Modeling

- What is the Nature of Communication?
  - Explicit
  - Implicit

Figure 8.1. Adaptive aiding (Rouse, 1988b).

An interesting aspect of answering this question concerns whether the emphasis should be on adapting *to* the user or adapting *of* the user. Although it is often the case that users' needs and preferences should be accommodated, there are also situations in which overall performance can be enhanced by providing users with new skills and knowledge. The concepts of embedded training and intelligent tutoring systems—discussed in Chapter 9—provide a strong basis for pursuing the notion of adapting the user as well as the aid.

Who Does the Adapting? If viewed very narrowly, this question has only one possible answer: the aid adapts. However, from a broader perspective, the agent of adaptation can be the system designer, users, or the aid. It can reasonably be argued that a system designer always adapts a system to a class of users and tasks. Beyond such "static" adaptation, users often configure a system for themselves by, for example, adjusting the seat and mirrors, choosing autopilot modes, or requesting particular report formats. The aid is the appropriate agent of adaptation if design adaptations need to be refined and/or changed, and if users are unlikely to perceive the need or be able to execute these adaptations.

When Does Adaptation Occur? It may be possible to adapt off-line before operation. Alternatively, adaptation can occur on-line in anticipation of changing demands. Finally, adaptation can occur on-line in response to changes. Clearly, the "what," "who," and "when" questions interact in the sense that not all possible combinations of answers are feasible. For instance, it is difficult to imagine a designer being the agent of on-line adaptation to the time-varying state of a particular user.

What Methods of Adaptation Apply? There are three general methods for aiding a user. An aid can make a task easier, perform part of a task, and completely perform a task. These three methods can be termed *transformation, partitioning,* and *allocation,* respectively (Rouse and Rouse, 1983). As examples, many display enhancement techniques (e.g., filtering and smoothing) represent transformations. Display highlighting (e.g., for cautions, alarms, and warnings) is an example of partitioning. Autopilots represent aiding via allocation.

Although the distinctions among these three methods of adaptation are certainly not crisp, it is a reasonably straightforward matter to decide which method applies if one focuses on the implications for the role of users. With transformation, users still perform the task in question. With partitioning, users are still "in the loop" but are not the sole agents of action. With allocation, the computer is the only active agent for the task that has been allocated.

How Is Adaptation Done? This question does *not* concern how the aid performs a portion or all of a task. Rather, the issue is the basis for determining the need for transformation, partitioning, or allocation. There are basically two approaches to making this determination. One approach is *measurement,* directly in terms of performance decrements or changes of demands, or indirectly in terms of, for example, the aforementioned leading indicators.

The other approach is *modeling,* whereby predictions of intentions, resource availability, and performance can be used to trigger adaptation. This approach basically provides the agent of adaptation with "expectations," the violation of which results in at least more targeted monitoring and eventually some degree of adaptation. The model-based approach to adaptation is discussed in more detail in the next section, "Intelligent Interfaces."

It is important to emphasize that depending on the agent of adaptation (e.g., users vs. aid), measurement and modeling have to be handled quite differently. If the user is to make the necessary measurements, the requisite information must be available and displayed appropriately. Furthermore, if the user is to have the necessary "mental models" to form appropriate expectations, it is likely that specialized training will have to be developed (and perhaps embedded in the system) as well as associated displays for supporting the use of these models. In contrast, if the aid is the agent of adaptation, measurement and modeling must focus on instrumentation and processing issues rather than displays and training.

What Is the Nature of Communication? The basic issue here concerns whether communication *about adaptation* should be explicit or implicit. Explicit communication between user and aid concerning the activities, awareness, and intentions of each party has the advantage of being minimally ambiguous. However, explicit communication can impose substantial overhead. The cost of this overhead can potentially exceed the benefits of aiding.

In contrast, implicit communication, via measurements and/or models, can greatly lessen this overhead, but suffers from greater uncertainty and ambiguity regarding the actions and intentions of each party. The trade-off between explicit and implicit communication hinges on the uncertainty associated with model-based implicit communication.

If a model can provide perfect predictions of the user's intentions and actions, there is no need to communicate explicitly. Thus, the cost of explicit communication can be avoided. However, as uncertainty grows, predictions will more frequently be wrong and, as a result, tasks will slip through the cracks or receive redundant efforts. To avoid these possibilities, increased explicit communication is needed to check or calibrate a model's predictions. If the level of uncertainty associated with a model's predictions becomes too great, totally explicit communication becomes the best policy.

---

1. The need for aiding can depend on the interaction of impending and recently completed task demands—task allocation decisions should not be based solely on the demands of the task in question.

2. The availability of aiding and who does the adapting can affect performance when the aid is not in use—total system performance may be enhanced by keeping the user in charge of allocation decisions.

3. When using measurements as a basis for adaptation, temporal patterns of user and system behavior can provide leading indicators of needs for aiding—it may be possible to use secondary indices as proxy measures of the indices of primary concern.

4. When using models as a basis for adaptation, the degree of task structure will dictate the accuracy with which inferences of activities, awareness, and intentions can be made—tasks with substantial levels of user discretion may limit the potential of model-based adaptation.

5. To the extent possible, incorporate within the aid models that allow predictions of the relative abilities of users and the aid to perform the task in particular situations—substantial variations of *relative* abilities of users and aids provide the central impetus for adaptation.

Figure 8.2. Principles of adaptation.

Design Principles. The data base of empirical studies of adaptive aiding is much too meager to codify a "principia adaptivia." However, sufficient R&D experience has been gained to be able to outline the general nature of the requisite principles and suggest specific design principles.

Two types of principles are needed: principles of adaptation and principles of interaction. Principles of adaptation concern when and how adaptive aiding applies, as well as the underlying mechanisms of adaptation. The experimental results noted earlier, and elaborated in Rouse (1988b), suggest the (admittedly incomplete) set of principles shown in Figure 8.2.

Interpretation of these principles is straightforward based on earlier discussions in this section. An exception is the second principle, which deals with an issue not previously discussed. Succinctly, the availability of aiding can affect the whole job, not just the tasks aided, and not just when the aiding is in use. For example, Morris found that *unaided* task performance was improved by the availability of aiding as long as users had control over when aiding was invoked (Morris, Rouse, and Ward, 1988).

The principles of adaptation shown in Figure 8.2 are very qualitative. Nevertheless, they help to answer some of the design questions posed earlier. For example, answering the question of who does the adapting can be seen to depend on task structure, likely task sequences, and the extent to which appropriate measurement and modeling methods are available and computationally viable.

Principles of interaction, an (incomplete) set of which are shown in Figure 8.3, relate to the characteristics of adaptive aiding that foster (or hinder) humans' acceptance and utilization of these aids. These principles also concern the extent to which users must understand the functioning of adaptive aiding in order to utilize the aiding appropriately and determine whether or not it is functioning properly. Finally, these principles relate to the somewhat traditional human factors issues of the displays and controls associated explicitly with adaptive aiding.

Interpretation of the principles in Figure 8.3 is quite straightforward, with the possible exception of the fourth principle. The "hot potato" problem refers to the possibility of a task being passed back and forth between user and computer. This might occur if, for instance, adaptation is based on workload measurements. High user workload would result in tasks being allocated to the computer, which would result in lowering the user's workload. Sensing this decrease, the computer might allocate tasks back to the user. This would result in user workload increasing, tasks being

1. Users can perceive themselves as performing better than they actually do and may want an aid to be better than they are—as a result, an aid may have to be much better than users in order to be accepted.

2. Ensure that user-initiated adaptation is possible and appropriately supported, even if aid-initiated adaptation is the norm—ensure that users feel they are in charge even if they have delegated authority to the aid.

3. Provide means to avoid user confusion in reaction to aid-initiated adaptation and methods for the user to preempt adaptation—make it very clear whether human or computer is supposed to perform a particular task at a specific time and provide means for changing this allocation.

4. It appears that aid-initiated "off-loading" of the user and user-initiated recapturing of tasks is a viable means of avoiding "hot potato" trading of task responsibilities—this asymmetry may help to ensure that users will feel in charge of the overall system.

5. There is a trade-off between the predictive abilities (i.e., in terms of uncertainty reduction) of models of human performance and intent, and the way in which the explicit vs. implicit communication issue is resolved—the cost of explicit communication (e.g., workload and time required) should be compared with the cost of adaptation errors (i.e., misses and false alarms).

6. The extent to which users can be appropriate agents of adaptation may depend on their models of the functioning of the aid and themselves—adaptation of the user (e.g., via embedded training) is a viable approach for providing such models.

7. A variety of specific human factors principles for design of complex information systems appear to apply to the design of the displays and controls associated with adaptive aiding—see discussions and references in Chapter 5.

Figure 8.3. Principles of interaction.

reallocated, and so on. This problem can be avoided by adopting a simple principle: computers can take tasks, but they cannot give tasks.

Summary. The notion of adaptive aiding is much more evolutionary than revolutionary. User-initiated adaptation has long been the norm in aerospace systems (e.g., autopilots). There are also many everyday examples of humans adapting their automobiles and appliances. Therefore, the primary innovation of adaptive aiding is not adaptation per se, but the possibility of aid-initiated adaptation. The design questions, in conjunction with the principles of adaptation and interaction, provide the basis for developing human-centered adaptive aiding.

INTELLIGENT INTERFACES

The foregoing discussions about traditional approaches to aiding and adaptive aiding provide a basis for considering comprehensive support systems. The remainder of this chapter is devoted to discussions of three comprehensive support systems. In this section, the architecture of the human-system interface for the intelligent cockpit discussed in Chapters 3 and 4 is considered. In a subsequent section, a support system design methodology is presented and illustrated with two examples, one of which is the design information system also discussed in Chapters 3 and 4.

The adoption of a human-centered design philosophy, which emphasizes using computers to support rather than replace humans, has substantial implications for design of human–computer interfaces, especially within complex systems. Appropriate designs of displays, input devices, and dialogue structures are no longer sufficient. These elements of human–computer interfaces now must be designed as integrated subsystems within an overall architecture that includes intelligent management of information and tasks. This requires a comprehensive architecture for an intelligent interface that can rival or exceed in sophistication other intelligent software subsystems envisioned for complex systems.

This section outlines an architecture for an intelligent interface based on the above concepts (Rouse, Geddes, and Curry, 1987). This interface architecture is discussed on two levels. First, it is presented as a general conceptual design for supporting users within complex aerospace, power, process, and manufacturing systems. Second, this architecture is presented as a specific design for the intelligent cockpit whose early evolution was chronicled in Chapters 3 and 4.

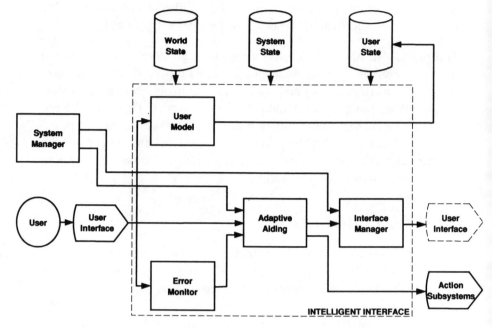

Figure 8.4. Overall architecture of intelligent interface.

Overall Architecture

A top-level view of the user–system architecture is shown in Figure 8.4. It is useful to discuss first those portions of this diagram that are not part of the interface. The system manager is responsible for coordinating the subsystems of the physical process of interest. For example, subsystems within the aviation domain might include those for various aspects of navigation, guidance, and control. Within the process control domain, subsystems might include inventory control, production monitoring, and energy management. The architectures of these subsystems are obviously very application dependent.

Figure 8.4, as well as subsequent illustrations, depicts the hardware associated with the interface (i.e., displays and input devices) as being external to the intelligent interface. Thus, the intelligent interface is manifested in software, which clearly emphasizes that the issues of primary concern are associated with human–computer interaction rather than traditional human factors. The purpose of making this distinction is not to denigrate concerns such as accessibility and life support, but to emphasize the fact that human-centered software design involves addressing and resolving many issues that are quite different than usual human factors considerations.

In addition, we have found that clarification of this distinction has usually increased the credibility of sophisticated interface concepts with traditional design engineers in the aerospace, marine, power, process, and manufacturing industries. For instance, our ability to work successfully in the military aviation community was greatly enhanced by being able to illustrate the importance of this distinction in the design of the intelligent cockpit (Rouse, Geddes, and Hammer, 1990).

The components in Figure 8.4 within the dashed lines comprise the general functionality of the intelligent interface. The architectures of these components can be described somewhat independently of application domain. Of course, detailed descriptions are necessarily very domain-dependent. The remainder of this section balances the general and specific levels of description by outlining the architectures of components in general and then providing specific examples from the intelligent cockpit application.

The functions depicted in Figure 8.4 serve two primary purposes—to overcome limitations and to enhance abilities. The purpose of the error monitor is to help users overcome their limitations by identifying and remediating human errors. From a complementary perspective, the purpose of adaptive aiding is to enhance users' abilities by flexibly supporting human performance in the manner described earlier in this chapter. The interface manager assists in overcoming limitations by managing the flow of information to the user. Without such assistance, this information flow might be overwhelming, considering the increasingly information-rich environments that usually result from implementing advanced information technologies in complex engineering systems. The interface manager also provides support that enhances abilities by adaptively representing and formatting information, which takes advantage of humans' perceptual strengths (e.g., recognition of complex patterns).

The level of sophistication of these types of support is highly related to capabilities of the user model to provide a real-time, dynamic knowledge base that enables assessing and predicting the user state. Several important trade-offs exist between the nature of the user model knowledge base and the extent to which the error monitor, adaptive aiding, and interface manager can support the user. The nature and impact of these trade-offs are discussed in the process of presenting the architecture of each of the components of the intelligent interface.

State Information

There are three types of state indicated in Figure 8.4: world, system, and user. These states are accessed in all of the figures in this section. Consequently, it is important to define their purposes at the outset. Succinctly, current and projected states of the world, system, and user are employed to ask "what is" (e.g., estimated current state) and "what if" questions (e.g., predicted state).

The definition of *state* differs for world, system, and user. The state of the world includes information on external demands, weather, terrain, and other features. The state of the system includes its dynamic state, the modes and failure status of subsystems, and information on current and upcoming operational phases and applicable procedures. The state of the user includes his or her activities, awareness, intentions, resources, and performance, the meanings of which are discussed below.

Current and predicted states are produced by a combination of measurements and models. The architectures of the world and system models, which produce the respective state estimates, are not discussed here because they are very application-specific and depend on the particular functions that report to the system manager. On the other hand, the state estimate produced by the user model is only directly needed by the functions within the intelligent interface. The user model is later discussed in some detail.

The concept of user state is central to the functioning of the components of the intelligent interface. It is important, therefore, to define the elements of user state quite specifically. As noted above, these elements include

- Activities: What is the user currently doing and likely to be doing, based on his or her intentions?
- Awareness: What active task requirements is the user consciously aware of and unaware of?
- Intentions: What are (will be) the user's goals and plans?
- Resources: What human information processing and input–output resources are (will be) available?
- Performance: How well is the user doing and likely to be doing, based on predicted activities and resources?

The user model maintains and updates current and predicted estimates of the above elements of user state. Later discussion illustrates how this is accomplished using an intent model, a resource model, and a performance model. It is useful, though, to discuss first how estimates of user state are used by the interface manager. In this way, it can be seen that estimates of user state need only be "good enough" to provide the desired functionality within the intelligent interface.

Interface Manager

The architecture for the interface manager is shown in Figure 8.5. The interface manager receives messages and requests from the system manager and adaptive aiding, including messages for the user or requests for information from the user. The role of the interface manager is to manage the flow of messages and requests in order to utilize effectively the user interface (i.e., displays and input devices), as well as the user's information processing resources and input/output channels.

To illustrate this concept, several of our efforts in aerospace systems have involved the application of expert systems technology to support navigation, guidance, and control (in commercial aviation) or mission planning, tactics planning, and situation assessment (in military aviation).

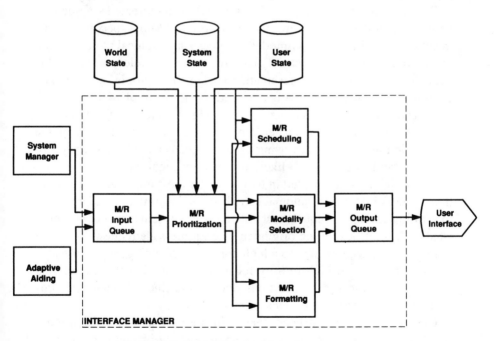

Figure 8.5. Architecture of interface manager.

Each of the expert systems involved requires a variety of information from the pilot and provides a plethora of potential messages for the pilot. Quite frankly, this type of support has the potential of being like water to a drowning man—the information has to be managed to be useful. This is the purpose of the interface manager.

The interface manager, therefore, is similar to an executive's assistant who zealously guards the supervisor's time and resources. This does not, of course, include preempting any user-initiated information transfer. In other words, the interface manager's control is unidirectional; demands upon the user are managed, but the user is free to initiate any activities of interest.

Prioritization and Scheduling. Some messages/requests are obviously more important than others. Importance can be expressed in terms of the criticality of the information in question, relative to achievement of one or more goals within the goal hierarchy of the domain of interest. An attribute of this criticality is the timeliness necessary for the information to be useful. If presented too soon, information may not yet be relevant; if presented too late, it may no longer be relevant.

Priorities are determined by the information requirements associated with successful pursuit of the user's intentions. More specifically, the intent model (within the user model) includes a hierarchical representation of the goals, plans, and actions that are potentially relevant within the domain of interest. This representation is annotated with information and control requirements that are determined using a mix of traditional task analysis methods and contemporary knowledge engineering techniques. In this way, the interface manager can obtain a baseline estimate of the information requirements by using the active goals and plans (elements of the user state) as pointers to likely information requirements.

This baseline is augmented in two ways. Of most importance, information requests by the user automatically become information requirements. In addition, messages/requests submitted via the system manager are annotated with priorities by the originator of each message/request. Thus, for example, a message related to a high-consequence emergency or threat is likely to be given a high priority regardless of the fact that the current assessment of the user's intentions does not include dealing with this event, perhaps because of a lack of awareness.

Once priorities are assigned, their utilization in scheduling is a fairly standard problem. There are various approaches to preemptive and nonpreemptive queue management that can be used to determine the order in which messages/requests will be serviced. It is not necessary to employ any sophisticated scheduling concepts or algorithms. This is due in part to the inherent inability of users to time-share in the same manner as computers—rapid, short duration sequencing of a large number of messages/requests is not reasonable.

Modality Selection and Formatting. Once messages/requests have been prioritized and scheduled, the interface manager has a resource allocation problem. The resources include the user's information processing capacity and input–output channels, as well as available displays and input devices. The demands for these resources are the prioritized messages/requests.

Resource requirements are characterized in two ways. First, requirements for human information processing and input–output resources are represented using a variation of Wickens's multiple resource theory (Wickens, 1984; Wickens, Tsang, and Pierce, 1985), which was discussed in Chapter 5. This multidimensional representation is the basis for the resource model (within the user model), which is discussed later in this section. Estimates of current and future available resources (elements of the user state) provide constraints within which modality selection and formatting must be resolved. Alternatively, these constraints can be violated by preempting current or intended activities. This might occur, for example, when the aforementioned high-consequence emergency or threat emerges.

The second way of characterizing resources involves available displays and input devices. The number and size of display surfaces, number and sophistication of audio channels, and number/modes of input devices limit the range of possible choices among modalities and formats. In general, however, the constraints imposed by human limitations are usually more restrictive.

Modality selection basically involves choosing among (as well as matching) resource channels for the three stages: (1) input to the user, (2) processing, and (3) output from the user. In general, this involves choosing displays for inputs, controls for outputs, and modalities and formats appropriate for the intervening human information processing.

These choices must take into account the current and projected availability of resource channels. Thus, using an aviation example, if the spatial–visual channel is currently devoted to a collision avoidance display where a potential near miss is being prompted, a lower-priority message that requires the spatial–visual channel (e.g., a hydraulic system status diagram) may be preempted while an even lower-priority message that requires the verbal–auditory channel (e.g., an approach clearance) may be transmitted without preemption. Alternatively, the lower priority hydraulic system message may be transformed to a simpler auditory status message that would have priority over the approach clearance.

The extent of resource competition can be lessened by suitable display formatting. This involves tailoring the nature of the presentation to the particular message/request and the user's current or most appropriate level of behavior. For instance, in situations where the user has time to be analytical (e.g., troubleshooting during cruise portions of a flight), displays might be quite different from those where an immediate, pattern-recognition-oriented response is necessary (e.g., responding to an engine fire or an approaching missile).

In general, formatting involves choosing message/request elements in terms of two dimensions: aggregation and abstraction (Rasmussen, 1986, 1988). The level of aggregation relates to the resolution of the message/request (e.g., quantitative vs. qualitative vs. status), which has direct implications for the size, labeling, and other features of message/request elements. The level of abstraction refers to the nature of the symbolic encoding of the message/request, ranging from physical form to functional purpose. The appropriate point in the aggregation–abstraction space depends on the intent of a particular message/request, as well as the user's current resources in terms of sensing, processing, and affecting. Therefore, for example, an aircraft pilot may only need a highly aggregated, abstract representation of another aircraft if that aircraft does not pose any immediate threat, either as an enemy or potential flight path conflict. But, in the event of an immediate threat, the pilot may want to know the type of aircraft and weapons, heading, speed, and altitude.

The dynamic allocation of human and system resources outlined here theoretically results in very efficient use of available resources. However, it potentially can result in considerable confusion on the part of a user. Unless constraints are added to the solution of the allocation problem, it is quite possible that a particular message, if considered at different points in time, could be presented in different locations, modalities, and formats.

As an aviation example, consider the implications of the aircraft attitude indicator being suddenly shifted from a visual to an auditory presentation because higher priority information demands the visual display surface. Then, owing to an urgent communication, attitude might suddenly be displayed via a tactile display. Clearly, frequent changes of this type could completely confuse a pilot. To minimize the possibility of these types of confusion, default or baseline locations, modalities, and formats are defined for the different types of message/request. Deviations from this baseline occur only if certain conditions are met. Further, to the extent that resource availability allows, these changes are prompted explicitly to assure that a user is aware of the deviations.

The current version of the interface manager within the intelligent cockpit has a predefined set of display pages, in terms of format but, obviously, not values of variables. While there is a default display for each information page, multiple display pages represent the same information at different points in the aggregation–abstraction space. In this way, elements on a display page are not independently manipulated within this space. Even though this limitation may be eliminated in future versions of the interface manager, it is clear that natural contextual relationships within a specific domain are such that complete freedom in formatting display elements (as opposed to pages) is neither necessary nor desirable.

To summarize, the interface manager is basically an on-line expert system for design of displays and programmable controls. The overall objective is adaptation to the information and control needs of the user. The processes depicted in Figure 8.5 provide, in effect, on-line task and workload analyses, with display modalities and formatting reflecting the results of these analyses.

Error Monitor

The architecture for the error monitor is shown in Figure 8.6. The error monitor reflects an especially important aspect of the human-centered design philosophy underlying the interface architecture. Namely, the goal is for the user to remain in control while computer-based systems provide support that helps the user to avoid negative consequences of inappropriate actions (Rouse and Morris, 1987).

Two complementary types of support are desirable. First, the frequency of user errors is decreased to the extent that such an approach does not cost too much in terms of lost human capability. For example, some types of interlock are useful, but excessive use of this method prevents the user from innovating when necessary.

The second type of support emphasizes avoiding unacceptable consequences without necessarily avoiding errors. By monitoring the user's

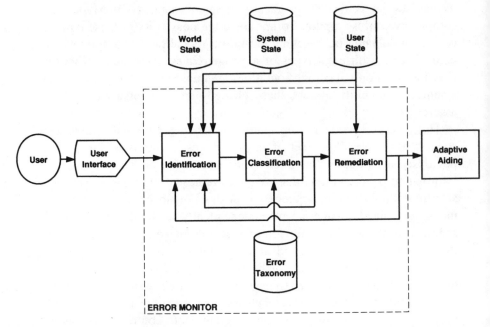

Figure 8.6. Architecture of error monitor.

behavior, relative to the states of the world and system, feedback can be provided that enables the user to detect anomalies quickly and reverse or compensate for consequences in a timely manner. In the extreme, the error monitor might recommend to adaptive aiding that control be allocated to an automatic system in order to avert impending high-cost consequences.

It is important to emphasize that the sophistication of the error monitor can range from quite simple—for example, the flight information system discussed in Chapter 7—to the full level of intelligence outlined in this section. While a lack of models and knowledge may preclude attaining the level of sophistication described here, an error monitor nevertheless can provide important support for overcoming users' limitations. More specifically, even if the computer does not know what a user should do in a given situation (i.e., automation is not possible), it may be feasible to determine whether or not what the user is doing is consistent. To illustrate, in the aforementioned flight information system, the error monitor did not know if a particular procedure was appropriate. However, it could determine if the procedure chosen was being executed correctly and, via appropriate feedback, substantially improve pilots' performance.

Thus, an error monitor can be developed incrementally. In terms of the elements of Figure 8.6, a baseline error monitor must at least be able to identify errors of omission. Identification of errors of commission requires more sophistication. A baseline level of error classification can be quite elementary (e.g., low- vs. high-consequence omissions), whereas more sophisticated classification schemes can involve explanations of identified errors premised on contextual reasoning. The possible levels of error remediation depend on the sophistication of identification and classification, as well as the capabilities of adaptive aiding (i.e., remediation recommendations have to be potentially executable).

The primary point of these introductory comments on the error monitor is that the interface architecture can provide incremental levels of support, depending on the levels of intelligence built into identification, classification, and remediation. In addition, and equally important, the architecture is inherently extendable as better understanding is gained. This property is also central to adaptive aiding, and is illustrated in the discussion of that component of the intelligent interface.

Error Identification. Identification of errors involves correlating the histories of the user's behavior and system response with operational procedures and scripts to detect anomalies between expected and observed behaviors. Manual methods for identifying errors were discussed in Chapter 7. The error monitor, however, requires on-line error identification which dictates computer-based methods.

Several approaches to algorithmic identification have been proposed (e.g., Rouse, Rouse, and Hammer, 1982; Hammer, 1984; Knaeuper and Morris, 1984). The error monitor employs Hammer's general approach to identification, with substantial extensions to take advantage of the capabilities of the user model. In particular, the intent model provides assessments of current goals, plans, and scripts, which enable much richer identification and classification. For instance, rather than only identifying errors of omission and commission, the error monitor employs a set of "critics" to look for specific anomalies that map more easily to explanations (e.g., critics for right control/wrong position and wrong control/right position).

There are three difficult aspects of applying these methods. First, substantial knowledge engineering is needed to compile the information necessary for the intent model to determine expected behaviors. Second, the decision as to whether or not a specific anomaly is an error depends heavily on previous actions. Consequently, as shown in Figure 8.6, error identification is partially based on feedback of the error state from error classification and remediation. This feedback helps to identify errors that are logical consequences of previous errors and, hence, dictate different types of message (if any) to the user. For example, the choice of an inappropriate procedure is likely to lead to identification of several incorrect actions unless the common explanation is determined—with the common explanation, a single message may be sufficient to remediate the error.

The third difficulty concerns errors of commission rather than omission. For the latter, it is relatively straightforward to decide if an omitted action is erroneous if the environment is sufficiently well structured. A high degree of structure makes it possible to predict likely action sequences and detect anomalies.

In comparing different domains, it can be seen that aircraft and spacecraft are fairly well structured during more demanding flight segments. Manufacturing, power production, process control, and marine operations have an intermediate degree of structure. Domains such as command and control, and perhaps engineering design, have a relatively low degree of inherent structure. It would appear, therefore, that the greatest level of sophistication could be obtained with aerospace applications. This is probably true if one is limited to implicit communications with the user regarding his or her intentions. However, as noted in earlier discussion, less structured domains also tend to have less stringent time constraints. This allows for direct queries regarding intentions. Thus, employing users' intentions to guide error identification, as well as other types of support, is not inherently limited to relatively constrained domains.

Regardless of degree of structure, errors of commission are much more difficult since they may reflect intentions of the user that go beyond current procedures and scripts. Alternatively, such "extra" actions may be irrelevant and harmless. As is explained in the later discussion of the intent model, errors of commission can be dealt with to a certain extent on the basis of a very much expanded knowledge base that links extra actions to possible additional goals (Geddes, 1985, 1989). For those actions that cannot be explained in this way, predictive models are needed to estimate the likely consequences of the unexplained actions. Unacceptable consequences lead to remediation, as explained in the following text.

Error Classification. Once an anomaly has been detected and identified as an error, it is necessary to classify it if other than an undifferentiated "error message" is to be provided. As discussed in Chapter 7, there are a variety of schemes for classifying errors. As noted in the discussions of the case studies in that chapter, a well-designed classification scheme can greatly facilitate determination and implementation of suitable remediation strategies.

For the purposes of the intelligent interface, errors are classified in terms of causes, catalysts, and consequences. Identification of causes is not necessarily essential to determining how an error might be dealt with. However, an understanding of underlying causes can be useful. Some distinctions among causes are important. In particular, the difference between slips and mistakes, as discussed in Chapter 7, has important implications for error remediation.

Errors due to basic human factors incompatibilities (i.e., slips) tend to be relatively easy to handle because they do not reflect inappropriate intentions. In contrast, when errors occur owing to misunderstandings (i.e., mistakes), it is likely that the user will require an explanation before accepting the conclusion that his or her choice was erroneous. Hence, for example, flipping the correct switch in the wrong direction might be automatically corrected, but choosing an apparently inappropriate airport approach procedure usually would not be overridden without explicit communication with the pilot.

Catalysts are factors that aggravate error-likely situations where incompatibilities and/or misunderstandings are present. Mental workload is a general example of a catalyst; distraction due to verbal communication is a more specific illustration. By identifying catalysts, it may be possible to eliminate their aggravating effects. For example, excessive workload due to a high level of manual activity or preoccupation with low-level displays might be lessened by decreasing the task demands on the user via adaptive aiding.

Knowledge of the nature and magnitude of likely consequences is essential to effective remediation. Consequences are often straightforward to determine via a domain-specific error taxonomy (e.g., the likely consequences of extending the landing gear at 600 knots). In many situations, though, it is necessary to develop a consequence model that can produce state-dependent predictions of likely consequences. For example, one would like to be able to predict whether or not the consequences of an apparently irrelevant action will undermine current goals and plans.

One very important problem with error classification is the cost associated with misclassification. Because the "best fit" causes, catalysts, and consequences lead to remediation of some type, it is essential that mechanisms be provided to avoid inappropriate remediation. For instance, the intelligent interface should not override the user when he or she is innovating to deal with a novel system failure.

Of course, it can be quite difficult for the error monitor to discriminate innovations from errors. The user, however, should be able to perform this discrimination. The error monitor attempts to minimize the chances of inappropriate overrides by providing levels of "error advisory" that inform the user (via adaptive aiding) of detected anomalies and recommended courses of action.

Error Remediation. Once an error is identified and classified, there are three general types of remediation that can be pursued. The lowest level of remediation involves *monitoring,* looking for particular events or consequences that support or reject hypotheses and might trigger more active remediation. This type of remediation reflects a more active level of reasoning than that associated with passive monitoring of action sequences. More specifically, remediation at this level involves active exploration of alternative explanations and courses of action. By informing adaptive aiding of this exploration—that is, lowest level of remediation—it is quite possible that adaptive aiding, because of its wider purview of task demands, may recognize impending difficulties earlier and initiate higher levels of remediation prior to the error monitor reaching similar conclusions.

The next level of remediation is *feedback,* which involves providing messages regarding the identification and classification of an error, as well as perhaps advice on appropriate compensatory actions. At a minimum, this type of feedback includes traditional alerting and warning systems. However, when appropriate and possible, feedback is much more intelligent. Such feedback involves, for example, synthesis of multiple warnings into an integrated explanation and recommendation.

The highest level of remediation is *control,* whereby the error monitor recommends active prevention of the propagation of consequences and may also recommend automatic initiation of compensatory actions. This is the level of remediation where misclassification of errors is potentially most problematic. Obviously, it is crucial that the user not be inappropriately preempted from acting. This possibility leads again to the notion of error advisories prior to initiation of error control.

The feasibility of error remediation at the level of control depends totally on such control being technically feasible. Although many functions may be always automated and others never automated, there are many functions (e.g., flight control) where the feasibility of automation depends on the performance requirements of the particular situation (e.g., tactical engagement vs. landing). Thus, it is by no means necessary for automation to be sufficiently sophisticated to do everything in order for it to be occasionally useful to do something.

The most suitable level of error remediation depends on the relative benefits and costs of each level, in terms of both system-oriented consequences and behavioral implications. The nature of these benefits and costs is highly context-dependent. To an extent, benefits and costs can be characterized using a knowledge-based approach to capturing domain-specific expertise. However, it is unlikely that such static knowledge will be definitive. The dynamic inference of current intentions provided by the intent model can expand the range of applicability of such a knowledge source. Adaptive aiding employs this type of information, in conjunction with its broader purview, to act upon the remediation recommendations it receives from the error monitor.

Despite this capability, users will occasionally have to judge benefits and costs, at least implicitly, by preempting remediation. From this perspective, the user is always in charge. In some situations, initiation of remediation requires explicit approval. In other situations, remediation occurs automatically unless it is preempted by the user. In any case, it is quite straightforward to allow users to tailor the preconditions for different levels of remediation relative to their individual preferences.

Adaptive Aiding

The architecture for adaptive aiding is shown in Figure 8.7. As with the error monitor, the design of adaptive aiding reflects a very important aspect of the human-centered philosophy underlying the intelligent interface. Specifically, as emphasized throughout this book, the goal is for the user to remain in control and be provided with aiding that adapts to current needs and capabilities in order to utilize human and computer resources optimally and, thereby, enhance overall performance. In this way, the user

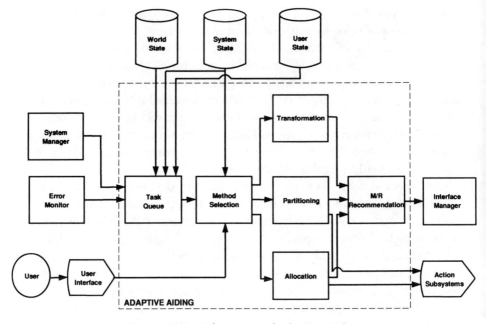

Figure 8.7. Architecture of adaptive aiding.

retains a maximum degree of control, but is not overwhelmed when the magnitude or nature of demands exceeds human abilities to perform acceptably.

Earlier in this chapter, the concept of adaptive aiding was introduced. In addition, a design framework was outlined and principles of adaptation and interaction presented. This section illustrates the role of adaptive aiding in the intelligent interface in general, and the intelligent cockpit in particular.

Task Queue. The task queue includes those tasks waiting to be done or projected as needing attention in the future. Based on world and system states, a standard or default set of tasks is determined as a function of the overall plan, operational phase, and situation. The user state, via the representation of the user's intentions, also provides indications of current and upcoming tasks. Additions (or deletions) to the set inferred from world, system, and user states are initiated by the error monitor and system manager.

Each task is dynamically classified in terms of who can potentially perform it: human only (H), computer only (C), or both (H/C). In addition, tasks have priorities and deadlines as well as contingencies that specify the preconditions for changes in priorities and deadlines. Further, there is information indicating relationships among tasks (e.g., complementary vs. competing).

There is some degree of latitude in defining tasks. For the research studies discussed earlier in this chapter, tasks included manual control, target recognition, and diagnostic procedure execution. In these cases, the time-varying demands of manual control, in combination with peaks and valleys in other task demands, made it appropriate to allocate target recognition or procedure execution to the computer.

It is quite possible to allocate entities that would normally be thought of as much larger in scope than a task. In the intelligent cockpit, allocations occur for scripts, tasks, and actions. Scripts are special cases of plans and include formal procedures and standard, but informal, task sequences for accomplishing particular goals (e.g., takeoff). Scripts may include tasks as elements (e.g., control nose wheel), and always include one or more actions (i.e., select, apply, release, etc.). Allocation of scripts is the most common form of adaptive aiding in the intelligent cockpit. Individual actions may be allocated as the result of error remediation recommendations from the error monitor. The following section on Method Selection discusses these as well as other forms of adaptive aiding.

Method Selection. As discussed earlier, there are three general methods for aiding a user: making a task easier, performing part of a task, and completely performing a task. The earlier discussion defined these alternatives as transformation, partitioning, and allocation, respectively, and illustrated the primary differences among these methods.

The choice among methods is constrained by the task designations defined above (H, C, and H/C). Aiding for H tasks can involve only transformation. Aiding for C tasks must, by definition, be in terms of allocation. Thus, method selection is automatic for H and C tasks. Although it might appear that this would allow "short-circuited" processing of these tasks, they are included in the queue to provide adaptive aiding with a purview of all tasks, as well as serve as a basis for message/request formulation regarding the status of H and C tasks.

The H/C tasks are amenable to the widest range of aiding. To be consistent with the human-centered design philosophy, the default allocation of H/C tasks is to the user. Tasks may be allocated to the computer if, at that point in time, any of the following are true:

- The user prefers to allocate the task to the computer (explicit user-initiated allocation),
- The user does not intend to perform the task (implicit user-initiated allocation),
- The user cannot perform the task acceptably (allocation due to direct inability),
- In order to perform the task acceptably, user performance on other tasks will be unacceptably degraded (allocation due to indirect inability), and
- The user is unaware of the task (allocation, perhaps preceded by prompting).

It is important that the user, at some previous time, has explicitly agreed to these rules.

The allocation of tasks proceeds as follows. Assume that the user has been performing all H and H/C tasks acceptably. Then, various cues (perhaps from the error monitor) start to indicate performance degradation or impending degradation. Adaptive aiding could respond to these indications by reallocating H/C tasks. A less drastic approach would be to try to improve human performance so as to eliminate the need for reallocation.

Initially, adaptive aiding would attempt transformation. For example, it might add prediction traces to the existing display elements for the task in question. As another example, a change of display modality might be attempted in order to decrease resource competition. If user performance improves as expected, then adaptive aiding would not initiate any additional aiding. However, if expectations are not met, adaptive aiding would attempt partitioning. As an aviation example, it might partition flight control into lateral and longitudinal control, and have the computer pick up the lateral control portion of the task. An illustration within process control might involve partitioning level and flow control within inventory management. If partitioning results in expected performance improvements, no further methods of aiding would be initiated. Otherwise, reallocation would be attempted.

This phased approach to adaptive aiding is appropriate when actual or predicted performance decrements are gradual rather than abrupt. Only then is it reasonable to do anything other than reallocation. More graphically, flirting with a performance "precipice" may be amenable to transformation or partitioning; floundering in a performance "abyss" requires allocation.

The phased approach is not symmetric in the human-to-computer and computer-to-human directions. Although adaptive aiding may decide on the transitions from transformation to partitioning, and partitioning to allocation, it is inappropriate for adaptive aiding to decide on transitions in the other direction. This necessary asymmetry reflects the principles of interaction discussed earlier (i.e., principle 4 in Fig. 8.3).

While it is generally reasonable for adaptive aiding to decide on transitions in the human-to-computer direction, there are situations in which the user should, if possible, make these decisions. As noted in the discussion of principles of adaptation (i.e., principle 2 in Fig. 8.2), the user's performance when *not* using an aid can be enhanced by being able to control when the aiding is employed. This possibility should be considered. But one should also be aware of those situations in which users cannot simultaneously be coordinators and performers, as well as situations in which users might utilize the aiding inappropriately.

Message/Request Formulation. As might be imagined, the functioning of adaptive aiding results in a need for much communication with the user. Adaptive aiding needs to monitor user state continually, particularly with regard to awareness and intentions relative to current and upcoming tasks. It also needs to communicate results of partitioning and allocation, at least to assure that everything is being done or monitored by some person or module.

It is important that the extent of these communications does not become excessive. Otherwise, user–aid interaction may become an overwhelming task in itself. The communication load can be decreased by either avoiding communication and accepting the resulting risks, or by communicating implicitly via models and unobtrusive measures. Alternative approaches and trade-offs associated with implicit communication were discussed earlier in this chapter.

The possibility of adaptive aiding transformations interacting with modality and format choices by the interface manager presents a potential problem. Although adaptive aiding dictates the content of the transformations, its recommendations for format may be inconsistent with the interface manager's overall perspective of available resources and competing objectives. Therefore, the "best" transformation may not be viable relative to the interface manager expeditiously scheduling it to appear via the user interface.

To avoid this problem, the interface manager must "understand" the intent of the messages/requests forwarded by adaptive aiding. As noted during the discussion of the interface manager, the most straightforward way to do this is to have the originator of a message/request specify its purpose (e.g., confirmation vs. routine update vs. unexpected update vs. warning vs. alarm). The need to specify the intent of messages/requests is an important requirement of all of the functions within the intelligent interface, as well as for the system manager.

User Model

From the foregoing discussion, it should be clear that the user model is central to the success of the other functions within the intelligent interface. As outlined earlier, the purpose of the user model is to estimate the current and projected user state in terms of activities, awareness, intentions, resources, and performance. This section discusses how this information is provided.

Figure 8.8 illustrates the architecture of the user model. Inputs to the user model include the user's sequence of actions for both information seeking and control, as well as the states of the world, system, and user. Outputs of the user model include updated estimates of current and projected states of the user.

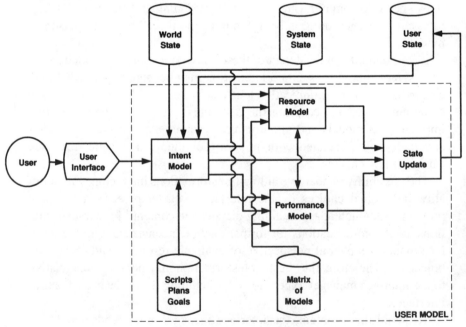

Figure 8.8. Architecture of user model.

Intent Model. The intent model interprets the user's actions, as well as the states of the world, system, and user, in the context of goals, plans, and scripts appropriate for the domain of interest (Geddes, 1985, 1989). An intention is defined as an active goal and the particular active plans and scripts in progress that support the goal. There are five basic steps of intent interpretation:

- Users' actions are decoded and passed to the script applier,
- If one or more scripts are matched, the actions are resolved and the active script is updated,
- If no match is found, the unresolved actions are passed to the plan inferencer,
- If one or more plans are found that are consistent with known goals, the actions are resolved and the scripts associated with these plans (if any) are activated, and
- If no match is found, the unresolved actions are passed to the error monitor.

This functionality is provided by two communicating modules, the script applier and the plan inferencer, each with its own knowledge base. The modules use a common representation of operator actions derived from Schank and Abelson's (1977) conceptual dependency representation for natural language understanding—the "language" to be understood in this case is users' keystrokes, button pushers, switch flips, and control movements. The script applier maintains an active set of scripts that have been activated by plans originating either in the a priori plan (e.g., mission plan or flight plan for aircraft operations, or production plan for manufac-turing operations) or inferred by the plan inferencer. These inferences result from applying reasoning mechanisms similar to those of Wilensky (1981) to a hierarchical knowledge base of goals, plans, and scripts.

This approach to intent interpretation has been found to work well for both fighter pilots and process control operators (Geddes, 1985, 1989). As might be expected, the accuracy and utility of the intent model's inferences are directly dependent on the degree of structure in the task environment. Fortunately, as noted earlier, it appears that loosely structured environ-ments do not have very stringent time constraints. This allows the in-telligent interface to query the user regarding intentions or, more likely, permits the use of a form of dialogue that inherently includes expressions of intentions. The intelligent interface can function quite well with either inferred or directly expressed intentions, or a mixture of both.

The outputs of the intent model include three components of user state: activities, awareness, and intentions. In the process of assessing and interpreting the user's intentions, the intent model also decodes button pushes, key presses, and control stick manipulations, and interprets these activities in terms of a standard set of action primitives that become elements of the user state. The awareness component of user state includes the active lists of goals, plans, and scripts.

Resource Model. The resource model estimates the user's current and projected resource utilization in terms of resources for input, central processing, and output. The user's activities and intentions enable the resource model to determine the current and projected task demands. These task demands are converted to resource requirements using Wickens' multiple resource theory (Wickens, 1984; Wickens, Tsang, and Pierce, 1985), with various heuristic refinements and extensions to compensate for the lack of data for many types of task and combinations of tasks.

Wickens' two-dimensional characterization of input resources is used, with one dimension being visual vs. auditory vs. tactile, and the other being spatial vs. verbal. Central processing resources are discriminated using the automatic vs. controlled characterization of Shiffrin and Schneider (1977), which is further decomposed using Rasmussen's (1986) skill-, rule-, and knowledge-based performance. Output resources are classified in terms of manual vs. vocal vs. visual.

In effect, the resource model maintains a human information processing "spreadsheet" of resource categories over time. The interface manager uses the projections on this spreadsheet to tailor and constrain the information flow to the user. Adaptive aiding uses this information as a basis for determining whether or not a task can be performed at all. If it can be performed, the next question is how well, which requires performance predictions. The performance model uses interactions among resource requirements as a factor in projecting user performance.

Performance Model. The performance model predicts the user's performance in current or potential future tasks. This information is essential to adaptive aiding if tasks are to be transformed, partitioned, and allocated appropriately. In earlier discussions in this chapter, a variety of models were noted as being useful for several part-task applications in aviation and process control. The intelligent interface requires a more comprehensive approach.

The idea of a comprehensive model of user behavior and performance is certainly far from novel—the literature is replete with a variety of attempts in this direction (Elkind et al., 1989; McMillan et al., 1989; Baron and Kruser, 1990). These types of "megamodels" present several difficulties that preclude them from being viable within the intelligent interface architecture. First, they tend to be unwieldy, requiring substantial computational resources for even fairly modest problems. Second, such megamodels can be rather obscure, requiring a high level of expertise and considerable insight in order to modify and extend the software. Finally, and certainly most important, the nature of these models is such that they are usually oriented toward answering a single type of question. The interface manager, error monitor, and adaptive aiding need answers to a wide variety of types of questions.

It appears that these difficulties are surmountable if the aforementioned "matrix of models" approach is adopted. The two dimensions of the matrix are types of tasks and types of performance metrics. Types of tasks include, at the highest level, discrete vs. continuous tasks, which are in turn decomposed into scanning, recognizing, problem solving, regulating, and steering. Types of performance metrics include the general categories of speed and accuracy, which are decomposed to more specific metrics for particular types of tasks.

The elements of the matrix are human performance models. Several versions of such a matrix have been developed previously for aviation (Rouse, 1981), process control (Rouse et al., 1982), and problem solving in general (Rouse, 1983). These compilations, when elaborated using modeling texts such as cited in Chapter 5, provide a necessary starting point for the matrix of models. But this set of models is not sufficient to provide the functionality required of the performance model.

The difficulty is that most of these models are oriented toward abstracted portions of tasks rather than the contextual reality within which these tasks actually occur. Many of the models discussed in Chapter 5 provide rather context-free representations of human performance in tasks that are highly relevant to many operational domains. However, few of these models capture the context-specific pattern recognition that dominates much of human behavior.

To overcome this limitation, the relatively context-free approach embodied in human performance models is combined with highly context-specific expert systems formulations. Thus, the matrix of models is part, but not all, of the performance model's knowledge source. The remainder of the knowledge consists of heuristics that specify when and how particular models apply, as well as appropriate parameter values for the applicable models as a function of resource availability and competition.

An especially important purpose of these heuristics is to enable predictions of multitask performance using a set of single-task or part-task models. It has not yet been possible, and may never be possible, to gather sufficient multiple-task performance data to allow empirical validation of all the possible and likely combinations of tasks of interest. As a result, opinions of users, human factors specialists, and others, as well as informal tests with part-task simulators, have to be used for augmenting the meager multiple-task data base available.

Considering the somewhat ad hoc nature of this "performance engineering" process, it is fortunate that the performance predictions of interest need not always be highly refined. Qualitative predictions of types of behavior frequently are sufficient. For instance, in some situations, the interface manager and adaptive aiding only need to know that the user is "overloaded." Then, the interface manager knows that further messages/requests should be inhibited, and adaptive aiding knows that partitioning or allocation should be considered.

A particularly good illustration of the sufficiency of qualitative estimates is in the area of performance criteria (i.e., how well the tasks are being performed). It appears that only four levels of performance assessment are needed:

- Performance is better than required (e.g., the user can probably handle more tasks, or at least more messages/requests),
- Performance is as required (e.g., new tasks or messages/requests are acceptable as replacements but not additions),
- Performance is less than required (e.g., performance enhancement is needed), and
- Performance is unacceptable (e.g., reallocation of tasks is needed).

It is important to emphasize that these estimates are qualitative, but the points of reference are likely to be crisply quantitative. Therefore, for example, altitude must be greater than 2000 feet or pressure must be less than 200 psi. However, the fact that altitude is greater or pressure is less than the reference points can be expressed qualitatively.

It is important that qualitative assessments and predictions are often sufficient. This allows for a much looser structuring of the performance model. Quantitative models can be used when available and warranted. Otherwise, qualitative approaches can be employed. Using a mix of quantitative and qualitative models in this way helps to avoid some of the difficulties of the megamodels discussed earlier.

Current Status of the Intelligent Cockpit

The comprehensive interface architecture outlined in this section may seem rather "blue sky" and difficult to imagine working. However, two pieces of evidence support the feasibility of the concept. The first piece of evidence was provided by an analytical assessment of the practicality and potential utility of implementing the adaptive aiding portion of the interface in the McDonnell-Douglas F/A-18 fighter aircraft (Geddes, 1986).

Three especially interesting conclusions were reached. First, the hardware and software architecture of the F/A-18 avionics already supports elementary interface management and could be extended to enable implementation of concepts such as discussed in this section. Second, the type of aiding most needed by the F/A-18 pilot involves planning and analysis. Finally, and most important, the adaptive aiding framework is not a major redefinition of the nature of aiding aircraft pilots. Instead, it is an approach to integrating automation that allows incremental inclusion of new capabilities without any fundamental change of the relationship between pilot and aircraft.

The second piece of evidence supporting the feasibility of the intelligent interface concept is the embodiment of this architecture within Lockheed's Pilot's Associate (Rouse, Geddes, and Hammer, 1990). The pilot–vehicle interface for the Pilot's Associate has evolved through nine prototypes. Initial emphasis was on realizing level A; subsequent efforts focused on moving toward level B within the evolutionary architecture (see Fig. 5.3) outlined in this section. As of this writing, effort is now directed at the goal of getting the entire system to run in real time in a procedural language on state-of-the-art avionics processors. This phase is a precursor to designing the Pilot's Associate into two aircraft, one an upgrade to an existing aircraft and another totally new aircraft.

DESIGN METHODOLOGY

In the past 20 or 30 years, there have been hundreds, if not thousands, of efforts to develop aiding and support systems. Very few of these systems have been truly successful. Furthermore, many of them have been rather ad hoc in the sense that while they seemed reasonable on the surface, at a deeper level there were few systems with consistent principled under-pinnings. This was particularly true for more comprehensive support systems.

From a human-centered design point of view, what is needed is a principled methodology that decreases the chances of ad hoc designs and, at the same time, takes advantage of the rich set of past experiences. In this section, such a methodology is discussed and illustrated with two examples, one from manufacturing and the other dealing with the design information system discussed in Chapters 3 and 4.

Alternative Support Concepts

In order to organize information on past experiences in developing aiding and support systems, more than 100 past aids and support systems were reviewed (Rouse and Rouse, 1983). Most of these previous efforts were in the aerospace industry. This review led to the conclusion that all of these efforts were concerned with supporting one or more of the set of 13 general tasks shown in Figure 8.9. This set of tasks was sufficient to classify and describe all of the aids and support systems reviewed in this analysis. A subsequent analysis of aids and support systems in the process control domain confirmed the general applicability of the scheme in Figure 8.9.

Two characteristics of the tasks in Figure 8.9 are of particular significance. First, most of the tasks involve generation, evaluation, and selection among alternatives. Use of this standard terminology is quite helpful for identifying approaches to enhancing abilities and overcoming limitations in these tasks. The second noteworthy characteristic of Figure 8.9 is the emphasis on alternatives in terms of interpretations of deviations, information sources, explanations, and courses of action. Hence, the structure of Figure 8.9 is based on a three-by-four array of action words (generation, evaluation, and selection) vs. objects of actions (types of alternatives). This degree of structure brings an important consistency and rigor to the process of characterizing user–system tasks.

- Execution and Monitoring
    1. Implementation of Plan
    2. Observation of Consequences
    3. Evaluation of Deviations from Expectations
    4. Selection Between Acceptance and Rejection
- Situation Assessment: Information Seeking
    5. Generation/Identification of Alternative Information Sources
    6. Evaluation of Alternative Information Sources
    7. Selection Among Alternative Information Sources
- Situation Assessment: Explanation
    8. Generation of Alternative Explanations
    9. Evaluation of Alternative Explanations
    10. Selection Among Alternative Explanations
- Planning and Commitment
    11. Generation of Alternative Courses of Action
    12. Evaluation of Alternative Courses of Action
    13. Selection Among Alternative Courses of Action

Figure 8.9. General set of user–system tasks.

Aiding requirements can be expressed in terms of needs to support one or more of the tasks in Figure 8.9. Equivalently, support requirements can be stated in terms of needs to assist users in generation, evaluation, or selection, and possibly implementation or observation. After reviewing hundreds of support system development and evaluation projects, it was concluded that there are basically 17 alternative ways to support users in performing the 13 general tasks in Figure 8.9. These 17 alternatives are summarized in Figures 8.10*a–e* and discussed below.

The most difficult support to provide is for generating alternatives, and there are few previous efforts to draw upon. This appears to be due, for the most part, to humans having great difficulty in specifying the attributes of desirable alternatives. Some progress has been made in using pattern

1. For a given situation, a support system can retrieve previously relevant and useful alternatives.
2. For a given set of attributes, a support system can retrieve candidate alternatives with these attributes.
3. Given user's assessments of suggested alternatives (e.g., via ranking or ratings), a support system can adapt its search strategy (e.g., attribute weights or logical operations) to produce new alternatives.

Figure 8.10*a*. Support concepts for generation of alternatives.

4. For a given alternative, a support system can assess the alternative's a priori characteristics such as relevance, information content, and resource requirements.
5. For a given situation and alternative, a support system can assess the degree of correspondence between situation and alternative.
6. For a given alternative, a support system can assess (e.g., via simulation) the likely future consequences such as expected performance impact and resource requirements.
7. For given multiple alternatives, a support system can assess the relative merits of each alternative.
8. Given user's assessments of evaluation results (e.g., via requests for explanations), a support system can adapt its evaluations in terms of time horizon, statistical measures, etc.

Figure 8.10*b*. Support concepts for evaluation of alternatives.

recognition methods to infer attributes of desired alternatives from a set of examples. It is fairly straightforward to retrieve the examples if users can define them appropriately, for example, all of the flight directors for previous jet fighter aircraft, or all the types of robot manipulators currently available. In general, as shown in Figure 8.10*a*, there are three basic ways to support generation of alternatives.

9. For given criteria and set of evaluated alternatives, a support system can suggest (e.g., via optimization) the selection that yields the "best" allocation of human–system resources.

10. Given user's assessments of selections (e.g., via ranking or ratings), reflecting perhaps individual differences and time-variations of criteria, preferences, and evaluations, a support system can adapt (e.g., by modifying criteria weights) its processing to provide suggestions that respond to these variations.

Figure 8.10*c*. Support concepts for selection among alternatives.

11. For given information, a support system can transform, format, and code the information to enhance human abilities and overcome human limitations.

12. For a given set of evaluated information, a support system can filter and/or highlight the information to emphasize the most salient aspects of the information.

13. For a given sample of information, a support system can fit models to the information in order to integrate and interpolate within the sample.

14. For given user and system constraints, as well as individual differences, a support system can adapt transformations, models, etc. (e.g., modify what information is presented and how it is presented).

Figure 8.10*d*. Support concepts for observation.

Evaluation of alternatives is easy to understand in that this type of activity is common within engineering analysis. The feasibility of supporting evaluation depends on the availability of appropriate models and calculation methods for the alternatives and measures of interest. The availability of models and methods can present difficulties when the phenomena of interest are complex.

15. For a given plan and information regarding the user's actions, a support system can monitor plan implementation for inconsistencies and errors of omission and commission.

16. For a given plan and information regarding the user's actions and intentions, a support system can perform some or all of the plan implementation to compensate for the user's inconsistencies, errors, or lack of resources.

17. Given information on intentions, resources available, priorities, etc., a support system can adapt its monitoring and/or implementation to reflect, for example, a change in goals.

Figure 8.10*e*. Support concepts for implementation.

Predictor displays are good examples of evaluative support for operators. Finite difference methods and geometric modeling techniques represent evaluative supports for designers. Spreadsheet models provide similar support for managers. As shown in Figure 8.10*b*, five approaches to supporting evaluation tasks encompass the range of alternatives that have been pursued in a variety of domains.

The majority of previous efforts to develop support systems have focused on selection among alternatives, in part because this type of support is most tractable. The two primary approaches to support are shown in Figure 8.10*c*.

If all of the alternatives have been specified, and the probability distributions associated with the consequences of choosing each alternative are known, and the users' criteria can be assessed, then it is usually quite easy to determine the best or optimal alternative. For alternatives involving multiple stages, locations, and so on, this optimization problem can become a bit tricky, but is, nonetheless, standard fare for control theory and operations research. This is not meant to denigrate the important function of selection support systems, but it is frequently found that identification of feasible alternatives and their likely consequences is sufficient for users to choose immediately without resorting to optimization. Thus, despite the great attention selection among alternatives has received, this task is usually not the most difficult task faced by users. Good support for generation and evaluation is typically more important, but also seldom available.

While the support of generation, evaluation, and selection is central to designing aiding and support systems, these types of support are not sufficient for a comprehensive approach to human-centered design. It can also be necessary to support implementation and observation. From a system design point of view, supporting observation involves display design, which has long been the stock-in-trade of human factors engineers. Fairly recently, methods based on expert systems technology have been developed for on-line, intelligent information management, a good example of which was discussed earlier in this chapter. These methods have the potential to filter, transform, and format information automatically. This capability can, for instance, make it feasible for users within complex systems to cope with the enormous amounts of data available via information technology. In general, as shown in Figure 8.10d, there are four ways to support observation tasks.

On the implementation (or action) side, it is quite feasible to have a computer monitor action sequences for reasonableness and consistency. An example of this type of support is the onboard information system for aircraft discussed in Chapter 7. A more comprehensive example is the intelligent cockpit, the interface architecture for which was discussed earlier in this chapter. The ideas embodied in these support systems can be generalized to yield the alternative approaches shown in Figure 8.10e.

All of the vast support system literature describes theories, design concepts, and evaluative results that relate to one or more of the capabilities summarized in Figure 8.10a–e. Support concepts for selection and observation are the most common; concepts for evaluation and implementation are not uncommon; and concepts for generation are fairly rare.

Eight-Step Design Methodology

The information in Figures 8.9 and 8.10a–e is accessed and utilized in the context of a structured methodology for determining the required functions in a support system. The eight steps of this methodology are indicated in Figure 8.11. In this section these steps are described. In two subsequent sections, two applications of the methodology are discussed.

Step 1: Define Tasks. This step is concerned with identifying the tasks to be supported. This is accomplished by analyzing the nature of the user's domain and determining what tasks are performed and how they are typically performed.

Step 2: Map to General Tasks. Use of this design methodology results in identification of the support functionality necessary for the tasks specified in step 1. As noted above, alterative support concepts are indexed in terms of one or more of the 13 general tasks in Figure 8.9. Since the set of general tasks is the intervening mechanism for identifying support

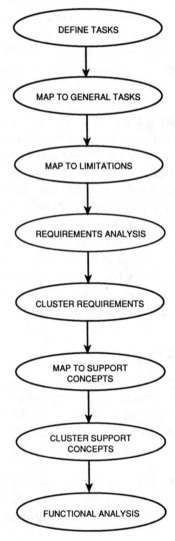

Figure 8.11. Design methodology.

concepts, this second step of the methodology involves mapping or linking each one of the tasks specified in Step 1 to one or more of the 13 general tasks. Each link is annotated with the reason for the link, as well as an application-specific interpretation of the connection.

Step 3: Map to Limitations. The next step involves considering the limitations that users are likely to face in performing the tasks specified by the links. Limitations can be characterized in more than one way.

Sage (1981) and Zachary (1986) have considered the *psychological limitations* associated with decision making and problem solving in complex systems. The difficulty with these tabulations of psychological limitations is that they are very general. With a little stretching, it is easy to imagine mapping each of the tasks emerging from step 2 to all of the psychological limitations. Such a result is not useful.

We have found it more useful to think in terms of *task limitations* that are tailored to the context of interest. For example, Figure 8.12 illustrates the types of task limitations relevant to information seeking and utilization in design. These limitations focus on getting things and doing things. For other applications, such as maintenance and operations, limitations might relate more to access of information and execution of procedures—in this case a different taxonomy than shown in Figure 8.12 would be needed.

- Not knowing *about* objects
- Not knowing *where* to get objects
- Not knowing *why* to get objects
- Not knowing *how* to get objects
- Not being *able* to get objects

- Not knowing *what* to do
- Not knowing *when* to do it
- Not knowing *why* to do it
- Not knowing *how* to do it
- Not being *able* to do it

Figure 8.12. Example of task limitations.

To perform this step, each of the tasks, actually specific-general task pairs, are reviewed to determine whether or not each of the limitations is likely to affect the task. This results in a number of task-limitation pairs that represent potential needs for support.

Step 4: Requirements Analysis. Each of the task-limitations pairs is then analyzed to determine in a very application-specific manner what is needed to overcome the limitation. This is accomplished by reviewing each task-limitation pair in terms of nature of the limitation, general task involved, and specific task involved. The utility of the results of this analysis can be enhanced by creating a computer-readable data base describing the requirements to overcome each relevant combination of limitations, general tasks, and specific tasks.

Step 5: Cluster Requirements. The set of task-limitation pairs, including the associated context-specific interpretations, are sorted into clusters with common limitations and general task attributes. This type of sorting is needed in order to map to the support concepts in Figures 8.10*a–e* which, as noted earlier, are indexed by general tasks. Limitations are used as a second sorting attribute because the interpretation of support concepts is influenced by the nature of the limitation that the support is to help overcome. The result of this process is clusters of task-limitation pairs that serve as input to the next step.

Step 6: Map to Support Concepts. Each of the clusters is then mapped to one or more of the general support concepts in Figures 8.10*a–e*. The mapping is determined by the nature of the general task associated with the cluster. Each relevant support concept is then interpreted within the context of the limitation associated with the cluster, as well as the application-specific tasks with links to the cluster.

This process can be substantially facilitated by developing and maintaining a highly structured and controlled vocabulary for expressing the specific instances of support needed. This greatly eases the burden of consistently and comprehensively managing the analysis process.

Step 7: Cluster Support Concepts. The data base of support concepts is then sorted using the terms of the aforementioned controlled vocabulary. The types of terms typically include

- Support verb: actions of the system to support the user,
- User verb: actions of the user that are enhanced by the support system's actions,
- Primary object: object of the user's actions, and
- Modifying object: noun plus preposition that modifies a primary object.

Typical entries in the database are of the form (support verb)(user verb)(modifying object)(primary object). An example entry in the data base for the design information system was: (Explain procedure)(to locate)(sources of)(drawing tools/packages).

Step 8: Functional Analysis. This step is concerned with converting the results of step 7 into a functional architecture for a support system. The purpose of the architecture is to provide the basis for realizing an integrated system that embodies the set of support concepts that emerged from step 6. Typical results of this functional analysis step are illustrated in the following example applications.

Application to Production Planning and Scheduling

Two very important characteristics of the design methodology outlined above are the underlying top-down, structured approach to conceptual design and the natural way in which aiding and support for multiple types of user can be designed. These characteristics are illustrated in this section via an example application of the methodology to production planning and scheduling (Rouse, 1988a).

Nature of Tasks (Step 1). Figure 8.13 presents a control-oriented perspective of production planning and scheduling, The terms facility, shop, and cell represent three of the five levels of the National Bureau of Standards' hierarchical control architecture (Ammons and McGinnis, 1986). These distinctions are particularly important in that different types of work occur on each level.

Industrial engineers work on the planning level and are concerned with optimal allocation of demand to facility resources (i.e., people, machines, and materials) in terms of release dates, due dates, and so forth. Shop supervisors work on the scheduling level and determine a daily schedule that is a modification of the production plan to accommodate rush orders and resource problems such as absent personnel, unavailable machines, and shortfalls or problems with materials. Finally, operations and maintenance personnel work on the operational level, making real-time sequencing and dispatching decisions and keeping machines in a productive state.

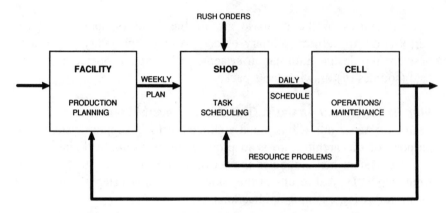

Figure 8.13. Production planning, scheduling, and operations.

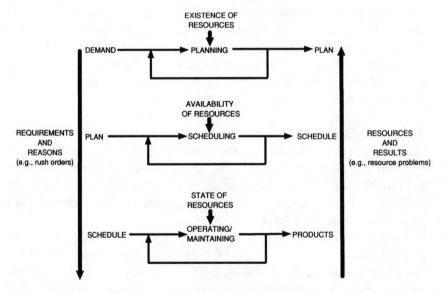

Figure 8.14. Hierachical view of production.

Figure 8.14 presents a slightly different view of production, with emphasis on the hierarchical relationships among levels. Two especially important distinctions are illustrated in this figure. While all three levels must deal with the same resources, they deal with them in very different ways. Planning focuses on the complement of resources, for the most part on an "as-designed" basis. Scheduling considers what resources are available at particular points in time. Operations and maintenance are concerned with, for example, whether or not particular machines are currently performing specific operations at an acceptable level of performance.

Another important distinction involves the ways in which information is shared among the levels in Figure 8.14. Rasmussen (1986) has noted that "looking up" in a hierarchy provides information on reasons for lower-level activities, whereas "looking down" provides information on resources available to pursue higher-level activities. Figure 8.14 expands on Rasmussen's notion by indicating also that information on requirements flows down while information on results flows up. The distinctions illustrated in Figures 8.13 and 8.14 have important implications for functional analysis and integration, which are discussed later in this section.

Approaches to Support (Step 6). The intermediate steps of the analysis for this example are summarized in Rouse (1988a)—in this chapter, these steps are illustrated in the discussion of the design information system in the next section. The remainder of the discussion of this production planning and scheduling example focuses on identifying support concepts and integrating them into a comprehensive support system architecture.

For production planning, support is needed for planning (general tasks 11–13 in Fig. 8.9) as well as the explanation aspects of situation assessment (tasks 9–10). For planning, supports for evaluating multiple alternatives (no. 7 in Figs. 8.10*a–e*) and selecting among alternatives on the basis of multiple criteria (no. 9) are appropriate. This type of support could be augmented, to an extent, by traditional materials resource planning methods. For explanation, support is needed for evaluating explanations emerging from the shop floor and integrating these multiple inputs into a facilitywide perspective (no. 5), as well as evaluating the resource consequences of accepting and acting upon particular explanations (no. 6).

The support requirements for shop scheduling are heavily oriented toward explanation (tasks 8–10), with some components of planning or real-time replanning (tasks 11–12). An important aspect of shop scheduling involves integrating situational information from multiple cells into an overall explanation of the shop situation. Support can be provided by mapping situational attributes to explanations that were previously germane (no. 1). If multiple explanations emerge, support would be needed to compare their "fit" and relative merits (no. 7). Once an explanation is selected, a support system can assist replanning by retrieving previously useful plans for working around whatever problems exist (no. 1). A support system can also help to assess the shop floor consequences of implementing a particular workaround plan (no. 6).

For cell operations, support is needed for both normal and off-normal operations (tasks 2–4). For normal operations, the support system can monitor action sequences to detect unmet expectations or inconsistencies (no. 15). If a planned course of action is available or otherwise evident, the support system can assist personnel in the execution of a plan in terms of, for example, flow control (no. 16). When a failure or other type of off-normal situation emerges, a support system can filter the typical plethora of symptoms (no. 12), as well as evaluate the credibility and consistency of available information (no. 4). A support system can also provide alternative explanations or diagnoses of symptoms (no. 1) and, in the event of multiple diagnoses, provide a comparison (no. 7).

Support System Integration (Steps 7-8). Figure 8.15 presents an overall support system architecture that was derived by connecting the inputs and output of the 14 approaches to support identified above. These 14 concepts were rather easily grouped into five high-level functions:

1. Facilitywide situation assessment,
2. Facilitywide planning,
3. Shop-level situation assessment and workaround planning,
4. Normal cell operations, and
5. Off-normal cell operations and maintenance.

Each block in Figure 8.15 represents a support function rather than an automated function. Accordingly, each block potentially involves user-system interaction via displays and input devices, design of which is initiated subsequent to use of the methodology illustrated in this section.

It is important to note that three types of user interact with this support system: facilitywide users at levels 1 and 2, shop floor users at level 3, and cell users at levels 4 and 5. For this reason, it is critical that the five levels

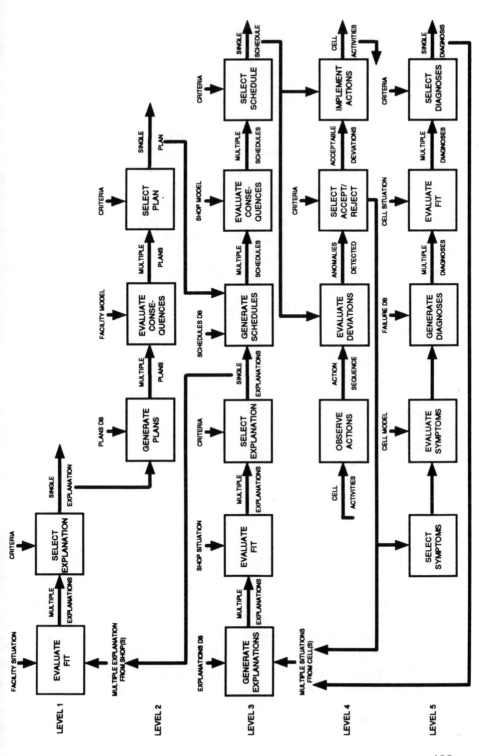

Figure 8.15. Overall support system architecture.

have a reasonably high degree of functional independence. Despite this requirement, information sharing and integration are still necessary.

Information sharing and integration can be considered in the context of Figure 8.14, where requirements and reasons flow down, while information on resources and results flows up. An important implication of this distinction for support system integration is the need to represent information flows at different levels of aggregation and abstraction depending on who is using the information (Rasmussen, 1986, 1988). From this perspective, integration involves much more than assuring plug-to-plug or word-to-word compatibility. Each of the above five functions has to process and display information in rather different ways. For example, while information on resources may reside in a single common data base, this information should be compiled, represented, and displayed quite differently for the three different types of users.

Another aspect of integrating the above five functions concerns the nature of what one function or level provides to another. In particular, higher-level strategies have to be consistent with feasible lower-level tactics. From this perspective, the nature of the weekly plan should be such that daily scheduling (e.g., workarounds owing to resource problems) is feasible without totally undermining the plan. Similar requirements exist for scheduling relative to operations. In general, the plans and schedules at all levels should allow for graceful departure from optimal. The type of integrated support system discussed in this section should help to achieve this objective. While such support does not inherently provide the requisite flexibility, the types of analysis outlined in this chapter should help to identify and deal with points in manufacturing control systems where such flexibility is crucial.

Application to the Design Information System

In Chapters 3 and 4, the results of the naturalist and marketing phases for the design information were discussed. In these discussions, it was noted that the results of these phases resulted in an hypothesis concerning designers' decision making and problem solving, within which the activities of information seeking and utilization occur. Central to this hypothesis is the three-dimensional space depicted in Figures 8.16 and 8.17.

Figure 8.16 depicts the abstraction and aggregation dimensions of the design space. This characterization is based on Rasmussen's constructs (1986, 1988). The definition of aggregation is obvious from the figure. Abstraction is more subtle.

LEVEL OF AGGREGATION	LEVEL OF ABSTRACTION		
	PURPOSE	FUNCTION	FORM
System			
Subsystem			
Assembly			
Component			

Figure 8.16. The abstraction and aggregation dimensions of the design space.

The concept of abstraction can be used to describe the types of representation relevant to design—representation provides the context within which information is sought and utilized. The three levels shown in Figure 8.16 can be defined as follows:

- Purpose: representation of design requirements, objectives to be met, problems to be solved, and so on, via scenarios, simulations, requirements documents, and so forth.
- Function: representation of relationships (i.e., physical, computational, etc.) via diagrams, equations, simulations, and so on.
- Form: representation of appearance (i.e., geometry, assembly, etc.) via drawings, pictures, mock-ups, and so forth.

Figure 8.17 depicts the task dimension of the design space. The tasks shown, and perhaps others that are similar in nature, can be viewed as designers' proximal intentions as they seek and utilize information.

Based on the three-dimensional design space depicted in Figures 8.16 and 8.17, the following hypothesis about design decision making and problem solving emerged: *Design behaviors can be characterized as trajectories or sequences of tasks in the design space that involve seeking and utilizing information while performing tasks at various levels of abstraction and aggregation.*

PURPOSE	FUNCTION	FORM
Explore Problem/Need	**Conceptualize Solution Functionality**	**Compose Form of Solution**
• Study current requirements (e.g., Statement of Work)– read and analyze	• Review functionality of past designs–read and analyze	• Review forms of past designs–read and analyze
• Study scenarios of operational need–view and analyze	• Synthesize/derive input-output relationships–create and represent	• Synthesize form of solution– create, visualize, and "sketch"
• Review requirements for past designs–read and analyze	• Develop model of functionality– integrate, analyze, and test	• Prototype/mock-up solution– integrate and fabricate
• Explicate performance attributes and criteria– integrate and decide	• Predict performance (exercise model)– calculate/simulate, analyze, and interpret	• Measure performance (collect data)–observe, measure, analyze, and interpret

Figure 8.17. The task dimension of the design space.

As discussed in Chapter 3, this hypothesis was tested by using trajectories (i.e., sequences of tasks) in the design space as a basis for creating design scenarios which designers were asked to rate along various dimensions (Rouse and Cody, 1989). The results supported the hypothesis, although the nature of the hypothesis makes it difficult to validate in a traditional scientific manner. Nevertheless, support was sufficient to use the design space as a basis for analyzing the requirements for design support and creating a functional architecture using the eight-step methodology (Rouse et al., 1990).

Define Tasks (Step 1). For this application, this step was completed by adopting the characterization of the design space. Thus, there were 12 tasks to be supported, each of which could occur at various levels of aggregation.

Map to General Tasks (Step 2). Two analysts independently performed the mapping from each of the 12 design tasks in Figure 8.17 to the 13 general tasks in Figure 8.17. Results were then compared and a consolidated mapping produced which had 64 links, or an average of somewhat over 5 links per design task. Each of these links was annotated with the reason for the link and a design-specific interpretation of the connection. This set of 64 design-specific instances of the general tasks served as the input to the next step.

Map to Limitations (Step 3). Each of the 64 tasks was reviewed to determine whether or not each of the 10 limitations in Figure 8.11 was likely to affect the task. A journeyman designer was assumed, which resulted in the conclusion that the "why" and "when" limitations in Figure 8.11 would not be relevant. Applying each of the remaining 7 limitations to the 64 tasks resulted in 213 task-limitation pairs that represented potential needs for support.

Requirements Analysis (Step 4). Each of the 213 requirements was then analyzed to determine in a very design-specific manner what needed to be done to overcome the limitation. This was accomplished by reviewing each of the 213 in terms of the nature of the limitation (Figure 8.11), general task involved (Figure 8.9), and design task involved (Figure 8.17). The results of this analysis formed a computer-readable data base describing the requirements to overcome each relevant combination of limitations, general tasks, and design tasks.

Cluster Requirements (Step 5). The 213 requirements, including the associated context-specific interpretations, were sorted into clusters with common limitations and general task attributes. As explained earlier, this type of sorting was needed in order to map to the list of support concepts (Figs. 8.10*a–e*) which is indexed by general tasks. Limitations were used as a second sorting attribute because the interpretation of support concepts is influenced by the nature of the limitation that the support is to help overcome. This process resulted in 43 clusters of requirements which served as input to the next step.

Map to Support Concepts (Step 6). Each of the 43 clusters was then mapped to one or more of the 17 general support concepts in Figs. 8.10*a–e*. The mapping was determined by the nature of the general task associated with the cluster. Each relevant support concept was then interpreted within the context of the limitation associated with the cluster, as well as the design tasks (Fig. 8.17) with links to this cluster. After several iterations, 613 instances of support resulted.

Cluster Support Concepts (Step 7). The data base of support concepts was then sorted using the terms of the controlled vocabulary explained earlier. Support verbs are summarized and defined in Figure 8.18. This information is the primary input to the next step, functional analysis. User verbs, modifying objects, and primary objects are listed in Figure 8.19.

SUPPORT	NUMBER	DEFINITION
Search	109	Use attributes or labels to identify and/or locate objects.
Execute	78	Perform procedures to assess, construct, evaluate, measure, etc.
Indicate	21	Display variables, relevant procedures, necessary activities, etc.
Transform	2	Modify, filter, and/or highlight observed or computed variables.
Explain	325	Interpret procedures, measures, variables, transforms, etc.
Tutor	78	Coach in use of procedures to access, construct, evaluate, etc.
Total	613	_____

Figure 8.18. Summary of support verbs.

The meaning of the words and phrases on these lists is quite straight-forward, with one exception. Both the singular and plural of drawings of form, model of functionality, and prototype appear as primary objects. The plural refers to retrieving relevant candidates, while the singular refers to creating a candidate.

It is interesting to note that with 6 support verbs, 11 user verbs, 8 modifying objects, and 36 primary objects, there are over 19,000 possible combinations. The 613 actual instances of support account for roughly 3 percent of these alternatives. Therefore, the results of the analysis process represent a much more structured and focused conclusion than a purely combinatoric tour de force would indicate.

Functional Analysis (Step 8). This step is concerned with converting the results of step 7 into a functional architecture for a design information system. The purpose of this architecture is to provide the basis for realizing the 613 instances of support that resulted from the requirements analysis. The overall architecture for the design information system is shown in Figure 8.20. Several aspects of this architecture are of special importance.

USER VERBS	MODIFYING OBJECTS	PRIMARY OBJECTS
Access	Availability of	Data collection plan
Construct	Between acceptance	Deviations of measured
Create	and rejection of	performance
Evaluate	Collection of	Deviations of predicted
Identify	Deviations from	performance
Locate	expectations of	Drawing tools/packages
Measure	Execution of	Drawings of form
Monitor	Explanations of	Drawings of forms
Obtain	Processing of	Experimental variables
Run	Sources of	Explanations of forms
Select		Explanations of functions
		Forms of past designs
		Functionality of past designs
		Information on forms
		Information on functions
		Information on operational needs
		Input/output representations
		Measured performance
		Model of functionality
		Models of functionality
		Model's predictions
		Model's variables
		Modeling tools/packages
		Off-the-shelf forms
		Off-the-shelf functions
		Off-the-shelf prototypes
		Past designs (forms)
		Past designs (functions)
		Past designs (requirements)
		Performance attributes and cri-
		teria
		Performance
		Performance data
		Prototype
		Prototypes/mock-ups
		Prototyping tools/packages
		Relevant input/output representa-
		tions
		Requirements information for
		current design
		Requirements information for
		past designs

Figure 8.19. Summary of user verbs, modifying objects, and primary objects.

First, any of the primary functions of search, execute, indicate, and transform can be accessed without having to access the other functions. Thus, the user is not constrained to following a particular design trajectory.

A second aspect of note is the ability to access the functions of explain and tutor from each primary function. In this way, explanations and tutoring can be tailored to the type of support being utilized and the current state of the design process. Note also that tutor can be accessed either via a primary support function or via explain. This provides the possibility of smoothly transitioning from an explanation that is not understood to more in-depth tutoring.

A third noteworthy aspect of Figure 8.20 is the interface, which appears twice in this diagram. The interface provides functionality that goes substantially beyond input devices and display pages. An intelligent interface such as discussed earlier in this chapter is envisioned.

The interface will have three central functions: information management, error monitoring, and adaptive aiding. The purpose of *information management* is to deal with the "overhead" associated with retrieving, creating, editing, and storing information in the process of utilizing the functionality shown in Figure 8.20. The role of *error monitoring* is to observe the data stream generated by the user, looking for inconsistencies and anomalies indicative of inadequate, inappropriate, or incorrect use of the functionality in Figure 8.20. The purpose of *adaptive aiding* is auto-

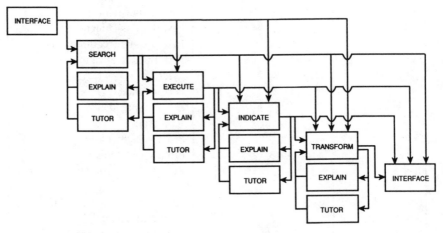

Figure 8.20. Overall architecture for design information system.

matic prompting, and occasionally initiation, of use of the functionality in Figure 8.20 to enhance information access and utilization in the design process.

As noted earlier, the key to providing the above type of support in an intelligent interface lies in being able to know what the human is doing and how these activities relate to goals and plans. This requirement can be satisfied for the design information system by recalling that the overall analysis is premised on design behavior being described as trajectories in the design space depicted in Figures 8.16 and 8.17.

If the intelligent interface knows where the user is in the design space, in terms of task and levels of abstraction and aggregation, as well as the likely trajectory of the user, then it should be quite feasible to provide the type of intelligent interface described above. In earlier discussion, it was indicated that much of a user's activities could be described, at least qualitatively, in terms of informal scripts and plans that emerge repeatedly and can provide the basis for inferring goals and plans. It seems reasonable to expect that "standard" trajectories in the design space can be identified and serve the same purpose (Rouse and Cody, 1989)

Current Status of the Design Information System. As of this writing, the design information system is in the conceptual design step of the engineering phase of the human-centered design process. Two activities are planned next. First, prototypes of the architecture in Figure 8.20 will be developed to embody the instances of support in Figures 8.18 and 8.19. These prototypes will be used to perform further marketing-oriented measurements. In parallel with this effort, detailed design will proceed on those aspects of the support system architecture that are found to include the most technical risk.

SUMMARY

This chapter has illustrated a wide range of alternative ways to aid users. Furthermore, a comprehensive methodology for design of aiding and support systems has been discussed. Applications in aviation, manufacturing, and engineering design have been described in detail. Thus, the methods, tools, and experience are available for providing the leverage that aiding offers.

However, aiding is not always sufficient. Moreover, sometimes the aiding needed is not technically feasible. Fortunately, training provides another leverage point that can replace aiding, compensate for aiding, or, more likely, complement aiding by providing a balanced approach to enhancing abilities and overcoming limitations.

REFERENCES

Adelman, L., Donnell, M. L., Phelps, R. H., and Patterson, J. F. (1982). An iterative Bayesian decision aid: Toward improving the user-aid and user-organization interfaces. *IEEE Transactions on Systems, Man, and Cybernetics, SMC-12*, 733–742.

Ammons, J. C., and McGinnis, L. F. (1986). Flexible manufacturing systems. In J. A. White (Ed.), *Production handbook* (Chapter 8.7). New York: Wiley.

Baron, S., and Kruser, D. S. (Eds.) (1990). *Human performance modeling.* Washington, DC: National Academy Press.

Chu, Y. Y., and Rouse, W. B. (1979). Adaptive allocation of decision making responsibility between human and computer in multi-task situations. *IEEE Transactions on Systems, Man, and Cybernetics, SMC-9*, 769–778.

Elkind, J. W., Card, S. K., Hochberg, J., and Huey, B. M. (Eds.) (1989). *Human performance models for computer-aided engineering.* Washington, DC: National Academy Press.

Enstrom, K. D., and Rouse, W. B. (1977). Real-time determination of how a human has allocated his attention between control and monitoring tasks. *IEEE Transactions on Systems, Man, and Cybernetics, SMC-7*, 153–161.

Forester, J. A. (1986). An assessment of variable format information presentation. In *Proceedings of NATO AGARD Symposium on Information Management and Decision Making in Advanced Airborne Weapon Systems.* Paris: NATO AGARD.

Freedy, A., Madni, A., and Samet, M. (1985). Adaptive user models: Methodology and applications in man-computer systems. In W. B. Rouse (Ed.), *Advances in man-machine systems research* (Vol. 2, pp. 249–293). Greenwich, CT: JAI Press.

Geddes, N. D. (1985). Intent inferencing using scripts and plans. *Proceedings of the First Annual Aerospace Applications of Artificial Intelligence Conference.* Wright-Patterson Air Force Base, OH: U.S. Air Force, pp. 160–172.

Geddes, N. D. (1986). *Opportunities for intelligent aiding in naval air-sea warfare: An A-18 war-at-sea study* (Rept. No. 8502-1). Norcross, GA: Search Technology, Inc.

Geddes, N. D. (1989). *Understanding intentions of human operators in complex systems.* Ph.D. dissertation, Georgia Institute of Technology.

Govindaraj, T., and Rouse, W. B. (1981). Modeling the human controller in environments that include continuous and discrete tasks. *IEEE Transactions on Systems, Man, and Cybernetics, SMC-11,* 410–417.

Greenstein, J. S., and Revesman, M. E. (1986). Development and validation of a mathematical model of human decision making for human-computer communication. *IEEE Transactions on Systems, Man, and Cybernetics, SMC-16,* 148–154.

Greenstein, J. S., and Rouse, W. B. (1982). A model of human decision making in multiple process monitoring situations. *IEEE Transactions on Systems, Man, and Cybernetics, SMC-12,* 182–193.

Hammer, J. M. (1984). An intelligent flight-management aid for procedure execution. *IEEE Transactions on Systems, Man, and Cybernetics, SMC-14,* 885–888.

Hammond, K. R., McClelland, G. H., and Mumpower, J. (1980). *Human judgment and decision making.* New York: Hemisphere/Praeger.

Knaeuper, A., and Morris, N. M. (1984). A model-based approach for on-line aiding and training in process control. *Proceedings of the 1984 IEEE International Conference on Systems, Man, and Cybernetics,* pp. 173–177.

Leal, A., and Pearl, J. (1977). An interactive program for conversational elicitation of decision structures. *IEEE Transactions on Systems, Man, and Cybernetics, SMC-7,* 368–376.

McMillan, G. R., Beevis, D., Salas, E., Strub, M. H., Sutton, R., and van Breda, L. (Eds.) (1989). *Applications of human performance models to system design.* New York: Plenum.

Morris, N. M., Rouse, W. B., and Ward, S. L. (1988). Studies of dynamic task allocation in an aerial search environment. *IEEE Transactions on Systems, Man, and Cybernetics, SMC-18,* 376–389.

Pearl, J., Leal, A., and Saleh, J. (1982). GODDESS: A goal-directed decision structuring system. *IEEE Transactions on Pattern Analysis and Machine Intelligence, PAMI-4,* 250–262.

Rasmussen, J. (1986). *Information processing and human-machine interaction: An approach to cognitive engineering.* New York: North-Holland.

Rasmussen, J. (1988). A cognitive engineering approach to the modeling of decision making and its organization in process control, emergency management, CAD/CAM, office systems, and library systems. In W. B. Rouse (Ed.), *Advances in man-machine systems research* (Vol. 4, pp. 165–243). Greenwich, CT: JAI Press.

Revesman, M. E., and Greenstein, J. S. (1986). Application of a mathematical model of human decision making for human-computer communication. *IEEE Transactions on Systems, Man, and Cybernetics, SMC-16,* 142-147.

Rouse, S. H., Rouse, W. B., and Hammer, J. M. (1982). Design and evaluation of an onboard computer-based information system for aircraft. *IEEE Transactions on Systems, Man, and Cybernetics, SMC-12,* 451–463.

Rouse, W. B. (1977). Human-computer interaction in multi-task situations. *IEEE Transactions on Systems, Man, and Cybernetics, SMC-7,* 384–392.

Rouse, W. B. (1981). Human-computer interaction in the control of dynamic systems. *Computing Surveys, 13,* 71–99.

Rouse, W. B. (1983). Models of human problem solving: Detection, diagnosis, and compensation for system failures. *Automatica, 19,* 613–625.

Rouse, W. B. (1986). Design and evaluation of computer-based decision support systems. In S. J. Andriole (Ed.), *Microcomputer decision support systems* (Chapter 11). Wellesley, MA: QED Information Systems.

Rouse, W. B. (1988a). Intelligent decision support for advanced manufacturing systems. *Manufacturing Review, 1,* 236–243.

Rouse, W. B. (1988b). Adaptive aiding for human/computer control. *Human Factors, 30,* 431–443.

Rouse, W. B., and Cody, W. J. (1989). A theory-based approach to supporting design decision making and problem solving. *Information and Decision Technologies, 15,* 291–306.

Rouse, W. B., Cody, W. J., Boff, K. R., and Frey, P. R. (1990). Information systems for supporting design of complex human-machine systems. In C. T. Leondes (Ed.), *Advances in Control and Dynamic Systems.* San Diego: Academic Press.

Rouse, W. B., Feehrer, C. E., Moray, N., Johnson, M. G., Nelson, W. R., Fragola, J., Rasmussen, J., Moran, T. P., Siegel, A. I., and Thorndike, P. W. (1982). Survey of development of models as applied to nuclear plant operators. *Proceedings of NRC Workshop on Cognitive Modeling of Nuclear Plant Control Room Operators.* Washington, DC: Nuclear Regulatory Commission, pp. 159–161.

Rouse, W. B., Geddes, N. D., and Curry, R. E. (1987). An architecture for intelligent interfaces: Outline of an approach to supporting operators of complex systems. *Human-Computer Interaction, 3,* 87–122.

Rouse, W. B., Geddes, N. D., and Hammer, J. M. (1990). Computer-aided fighter pilots. *IEEE Spectrum, 27,* 38–41.

Rouse, W. B., and Morris, N. M. (1987). Conceptual design of a human error tolerant interface. *Automatica, 22,* 231–235.

Rouse, W. B., and Rouse, S. H. (1983). *A framework for research on adaptive decision aids* (Rept. TR-83-082). Wright-Patterson Air Force Base, OH: Aerospace Medical Research Laboratory.

Sage, A. P. (1981). Organizational and behavioral considerations in the design of information systems and processes for planning and decision support. *IEEE Transactions on Systems, Man, and Cybernetics, SMC-11,* 640–678.

Schank, R. C., and Abelson, R. P. (1977). *Scripts, plans, goals, and understanding.* Hillsdale, NJ: Lawrence Erlbaum Associates, Inc.

Schneider, W., and Shiffrin, R. M. (1977). Controlled and automatic human information processing II: Perceptual learning, automatic attending, and a general theory. *Psychological Review, 84,* 127–190.

Walden, R. S., and Rouse, W. B. (1978). A queueing model of pilot decision making in a multi-task flight management situation. *IEEE Transactions on Systems, Man, and Cybernetics, SMC-8,* 867–875.

Wickens, C. D. (1984). Processing resources in attention. In R. Parasuraman and R. Davies (Eds.), *Varieties of attention.* New York: Academic.

Wickens, C. D., Tsang, P., and Pierce, B. (1985). The dynamics of resource allocation. In W. B. Rouse (Ed.), *Advances in man-machine systems research* (Vol. 2, pp. 1–49). Greenwich, CT: JAI Press.

Wilensky, R. (1981). PAM. In R. C. Schank and C. K. Riesbeck (Eds.), *Inside computer understanding* (pp. 136–179). Hillsdale, NJ: Lawrence Erlbaum Associates, Inc.

Zachary, W. (1986). A cognitively based functional taxonomy of decision support techniques. *Human-Computer Interaction, 2,* 25–63.

Chapter 9

Design of Training

Training is the process of managing people's experiences so that they gain the requisite knowledge and skills that give them the potential to perform. The extent to which this potential is created depends on the nature of the training experiences, as well as the aptitudes and abilities of the personnel being trained. Training is an investment in people.

In contrast, aiding focuses on augmenting task performance directly. Consequently, rather than emphasizing knowledge and skills as is the case with training, the emphasis in aiding is on task-specific behaviors to be prompted and enhanced. The means for prompting and enhancing behaviors usually involves technology, often information technology. Thus, aiding tends to be an investment in technology.

People vs. technology. How should one invest? While a book on human-centered design inevitably is biased a bit in the people direction, this question cannot be answered simply.

For example, the two types of investment are not readily comparable. Training involves downstream recurring costs, while aiding involves upstream capital costs. Further, the choice is not really between training *or* aiding. Instead, it is an issue of balance. How should training and aiding be integrated to best accomplish objectives?

This question is pursued in a later section of this chapter. First, it is necessary to discuss training at a level of detail commensurate with the treatment of aiding in Chapter 8. This will provide the basis for both designing training itself, and assessing alternative combinations of training and aiding.

DESIGN OF TRAINING SYSTEMS

The design of training includes a broader set of issues than does the design of aiding. This is due to the fact that training involves a set of activities and processes that, in themselves, constitute a system. In contrast, aiding is usually embodied in functionality that is added to an operational system. Therefore, a training system is more analogous to an operational system than it is to aiding. A training device—for example, a simulator—within a training system is a better analogy to aiding within an operational system. In this chapter, the design of both training systems and training devices is considered.

Instructional System Design

Training system design is often also called instructional system design (Gagne, Briggs, and Wager, 1988). Figure 9.1 lists seven steps of instructional design. This list was gleaned from a compilation of instructional design methods used in the military (Booz-Allen and Hamilton, 1985).

As might be expected, the design process begins with consideration of the tasks and duties of the personnel of interest. Of particular interest are those tasks whose performance must be certified. Good examples include many tasks of aircraft pilots and nuclear power plant operators. Often, training for tasks that require certification involves real equipment or simulators, which serve as the means for performance assessment and certification.

The second step is concerned with determining the knowledge and skills with which trainees enter the training system. In some cases, one can select the population of trainees so as to assure that high knowledge and skills prerequisites are satisfied. In other cases, the population of trainees is predetermined and the training system must accommodate them regardless of their a priori knowledge and skills.

Training requirements are next defined in terms of the additional knowledge and skills that trainees will have to gain in the training system in order to perform satisfactorily. Obviously, the greater knowledge and skills

1. Define the tasks and duties of personnel, particularly the tasks that must be certified through the use of training equipment or simulators.
2. Identify the prerequisite knowledge and skills found in the target population to be used in operating the system.
3. Define the training requirements by determining the additional training that will be needed by the training population.
4. Define the methods for training the required knowledge and skills.
5. Identify the training equipment and simulation required to support the training curriculum.
6. Prepare the course material.
7. Evaluate the course by presenting it formally.

Figure 9.1. Instructional system design.

people begin with, the less new knowledge and skills needed, and hence the less training needed.

Knowledge and skill requirements provide the basis for determining appropriate methods of training. The basic goal is to identify, locate, or create effective and efficient means for imparting the knowledge and skills needed.

The choice of training methods has strong implications for the training equipment and simulation needed to support and embody these methods. These devices also have to be integrated into the overall curriculum which, in many cases, will be a revision and/or addition rather than a totally new curriculum.

Next, the course materials are prepared. This includes the plan of instruction, syllabus, instructor lecture materials, and handout materials for trainees. Supporting materials for use of the training devices must also be prepared, including operating instructions, scenarios or problems to be used, and scoring methods.

Finally, the course is evaluated by formally presenting it. Formative testing, verification, and demonstrations should have preceded this summative evaluation. Hence, the primary issue of concern should be the extent to which the training system satisfies requirements, that is, provides the knowledge and skills needed. Unfortunately, the question of whether or not these knowledge and skills result in acceptable on-the-job performance—whether or not the training is valid—is seldom addressed. A later example illustrates how this question can be pursued.

Knowledge and Skill Requirements

We have found that the determination of knowledge and skill requirements, and mapping these requirements to training methods and devices, can be assisted by adopting a structured approach to this analysis (Johnson et al., 1985; Fath and Rouse, 1985). This approach views skill as the proficient application of knowledge gained by practice and repeated use. From this perspective, beyond the need for practice, the critical issue concerns knowledge requirements.

It is useful to characterize knowledge requirements in terms of operational knowledge and system knowledge. Operational knowledge (Fig. 9.2) denotes information about the way in which tasks are performed. System knowledge (Fig. 9.3) refers to information about the system within which operators, maintainers, and managers perform their tasks.

LEVEL	TYPES OF KNOWLEDGE		
	"WHAT"	"HOW"	"WHY"
Detailed/ Specific/ Concrete	Situations (What Might Happen)	Procedures (How to Deal With Specific Situations)	Operational Basis (Why Procedure is Acceptable)
	Criteria (What is Important)	Strategies (How to Deal With General Situations)	Logical Basis (Why Strategy is Consistent)
Global/ General/ Abstract	Analogies (What Similarities Exist)	Methodologies (How to Synthesize and Evaluate Alternatives)	Mathematical Principles/Theories (Why: Statistics, Logic, Etc.)

Figure 9.2. Types of operational knowledge.

The contrast between Figures 9.2 and 9.3 can be summarized by noting that operational knowledge concerns *how to work the system,* while system knowledge relates to *how the system works.* As discussed below, this distinction has important implications for choosing among training methods. This distinction does not, however, imply that operational knowledge and system knowledge are mutually exclusive.

In general, analysis of operational knowledge requirements leads to determination of system knowledge requirements. In other words, analysis of how to work the system leads to determining the extent to which personnel need to know how the system works. The results of such an analysis tend to be quite different depending on whether the personnel of interest are maintainers, operators, managers, or designers. For instance, knowledge of principles and theories tends to be more important for designers than maintainers.

Somewhat surprisingly, knowledge of principles and theories appears to have very little impact on the performance of most jobs. Extensive studies of both operations (Morris and Rouse, 1985a) and maintenance (Morris and Rouse, 1985b) have indicated that knowledge of standard procedures is a better predictor of good performance than is knowledge of theories and principles. The implication of these results is that inclusion of knowledge

LEVEL	TYPES OF KNOWLEDGE		
	"WHAT"	"HOW"	"WHY"
Detailed/ Specific/ Concrete	Characteristics of System Elements (What Element is)	Functioning of System Elements (How Element Works)	Requirements Fulfilled (Why Element is Needed)
↓	Relationships Among System Elements (What Connects to What)	Cofunctioning of System Elements (How Elements Work Together)	Objectives Supported (Why System is Needed)
Global/ General/ Abstract	Temporal Patterns of System Response (What Typically Happens)	Overall Mechanism of System Response (How Response is Generated))	Physical Principles/Theories (Why: Physics, Chemistry, Etc.)

Figure 9.3. Types of system knowledge.

and theories in training programs needs more justification than simply the traditional "It's good for them."

Training Methods

Figure 9.4 characterizes a range of approaches to training. The methods noted as "passive" in the table tend to be instructor-dominated in that the trainee passively consumes knowledge when these methods are used. In contrast, the "active" methods tend to be trainee-dominated in that the trainees actively utilize knowledge, hopefully in a manner that has been designed to clarify, reinforce, and extend operational and system knowledge. Thus, to a great extent, active training methods can be viewed as carefully planned surrogates for actual experience.

Figure 9.4 indicates that some methods are better for imparting system knowledge than operational knowledge, and vice versa. The classifications shown are obviously very approximate. However, these classifications are consistent with the traditional distinction between "in-the-head" and "in-the-hands" training. Moreover, as is shown below, these classifications emphasize the need to design training programs with significant "active" components.

EMPHASIS	METHODS		
	"PASSIVE" TRAINING	"ACTIVE" TRAINING	ACTUAL EXPERIENCE
System Knowledge	Classroom Lecture	Equipment Mock-ups	Induced Malfunctions
	Classroom Discussion	Flat Panel Simulators	Practice With Real Equipment
	Video and Films	Part-Task Simulators	On-the-Job Apprenticeship
Operational Knowledge	Laboratory Demonstrations	Full-Scope Simulators	On-the-Job Responsibility

Figure 9.4. Methods for acquisition of knowledge and skills.

The training methods in Figure 9.4 can be differentiated further on the basis of their relative effectiveness and efficiency. Effectiveness is defined as the degree to which a method can successfully support the acquisition and retention of the desired type of knowledge and/or skill. In other words, effectiveness relates to the satisfaction of knowledge and skill requirements. Therefore, effectiveness of training is defined in a similar manner to effectiveness of equipment designs, that is, relative to requirements.

Efficiency is defined in terms of the time and resources required to achieve success: to be effective. Efficiency can be low for several reasons. For example, it takes a very long time to infer system knowledge via active training methods—the process of discovery may require many trials. As another illustration, it takes a long time to gain both operational and system knowledge on the job because many other job-related activities interrupt the training process.

Figure 9.5 presents an assessment of the relative effectiveness and efficiency of the methods in Figure 9.4. Passive training methods are both effective and efficient for imparting all aspects of system knowledge. Active methods are effective, but inefficient for system knowledge. Actual experience is very inefficient.

PRODUCT	METHODS		
	"PASSIVE" TRAINING	"ACTIVE" TRAINING	ACTUAL EXPERIENCE
"What"/"How" (System)	Effective and Efficient	Effective but Inefficient	Effective but Very Inefficient
"What"/"How" (Operational)	Very Ineffective but Efficient	Effective and Efficient	Effective but Inefficient
"Why" (System)	Effective and Efficient	Effective but Inefficient	Ineffective and Very Inefficient
"Why" (Operational)	Ineffective but Efficient	Effective but Inefficient	Effective and Very Inefficient

Figure 9.5. Effectiveness and efficiency of alternative methods.

In terms of operational knowledge, passive methods are not effective. Active methods and actual experience are effective, but incur varying levels of inefficiency. Of course, the general inefficiency of actual experience is partially compensated for by the fact that personnel can do other useful things as they are gaining operational knowledge.

Training Programs

Figures 9.2 through 9.5 provide the basis for synthesizing a mix of training methods into an integrated and comprehensive training program. First, overall training objectives (task behaviors and/or performance requirements) are decomposed into operational knowledge requirements using Figure 9.2. For example, for an entry-level maintenance job, operational knowledge requirements are likely to be in the categories of situations, criteria, and procedures. The results of an analysis using Figure 9.2 would be lists of context-specific knowledge in each of these three categories.

The next step involves determining system knowledge requirements using Figure 9.3 as a guide. For the entry-level maintainer, system knowledge requirements are likely to be in the categories of characteristics and relationships, with perhaps some knowledge in the category of functioning. The overall assessment of operational and system knowledge requirements can be performed by instructional system developers and domain experts for jobs where much previous experience is available. For jobs where there is not a strong baseline, much engineering analysis may be needed to determine knowledge requirements.

The third step involves considering alternative training methods. The nature of the knowledge requirements is used to choose among methods based on the guidance in Figures 9.4 and 9.5. For the entry-level maintainer, operational knowledge requirements focus on "what" and "how," and tend to be more crucial than system knowledge requirements. Figure 9.5 indicates that active training methods are likely to be the best choice for operational knowledge. However, if efficiency is not a concern, actual experience may be substituted for active training methods. Passive training methods are a good choice for the system knowledge requirements.

Choosing among training methods within a category involves assessing the strengths and weaknesses of specific instances of each of the methods in Figure 9.4. Issues influencing choices include size of trainee population, availability of training devices and real equipment, and budgets available. Choices among training methods also must be based on the time and costs of developing, implementing, and supporting the alternative methods (Johnson et al., 1985).

The fourth step of the process involves integrating the mix of methods chosen into a coherent training program. This includes considering how different levels of training device fidelity can best be combined to achieve training objectives—discussed in the next section. Also central to this step are the analyses associated with developing an overall training system (e.g., Gagne, Briggs, and Wager, 1988). This requires consideration of a broader set of issues (e.g., personnel selection) than can be pursued in this chapter—see Booher (1990) for elaboration of these issues.

MIXED-FIDELITY APPROACH TO TRAINING

Over the last decade or so, simulation has emerged as a very popular approach to training. While this method has been used in the aircraft industry for many years, simulation is a more recent innovation in the marine, process, and power industries. Most training simulators involve a mock-up or facsimile of the equipment system of interest, with all the buttons, switches, meters, and gauges that are available on the real equipment. A computer is used to determine how temperatures, pressures, speeds, and so on, change in response to actions by the trainee.

A key issue in the use of simulators is the level of fidelity necessary to assure transfer of training from the simulator to real equipment. A wide variety of conceptualizations and definitions of "fidelity" are available (Su, 1984). A useful working definition of fidelity is the accuracy with which the simulator reproduces the appearance and behavior of the real equipment.

High-fidelity simulators can be very expensive. Consequently, it has often been suggested that rougher approximations to the real equipment might be adequate, at least when combined with on-the-job training. Unfortunately, it has proven difficult to determine which approximations are appropriate. As a result, a good part of the aviation, marine, process, and power industries has invested in high-fidelity simulators, typically at costs of $2,000,000 to $20,000,000 per simulator.

Problems With High-Fidelity Simulators

Unfortunately, the use of high-fidelity simulators presents three problems. First, the cost involved is usually too high to allow a training facility to have more than a single simulator. This constraint limits class size and substantially increases the cost per trainee. Alternatively, if cost-effective class sizes are maintained, each trainee spends very little time using the simulator or group exercises must be employed, which may limit the effectiveness of the experience for each trainee.

A second problem with high-fidelity simulators is due to the fact that many aspects of a particular simulator may replicate features of an equipment system that is significantly different from the system that the trainees will eventually operate, maintain, or manage. Whereas this is usually not the case in the aviation industry, it definitely is true for the marine, process, and power industries, where there is relatively little standardization among ships and plants. For these industries, complete reliance on high-fidelity simulators may result in a substantial portion of the trainees' time being devoted to learning specific features of an equipment system that will not be seen again once trainees are on the job. Clearly, this is a waste of a very expensive resource.

The third problem with high-fidelity simulators stems from the complexity they inherently have if they are faithful reproductions of complex real equipment systems. This complexity, especially in terms of the plethora of meters, gauges, alarms, and annunciators in a typical simulator, can make it difficult for beginning trainees to understand the fundamental processes occurring in the equipment system of interest, as well as the ways in which these processes are affected by different events and actions.

A Mixed-Fidelity Solution

A solution to these types of problems becomes apparent when one considers the roles that simulators should play in the process of acquiring system and operational knowledge. The results of such an analysis led to the concept of mixed-fidelity training (Rouse, 1982).

Figure 9.6 depicts the relationships among simulators that form the basis for mixed-fidelity training. The low-fidelity simulator typically bears little resemblance to the equipment system of interest. This is due to the fact that its purpose is not for learning about the equipment system, but instead is for learning basic skills and general principles that can later be transferred to real equipment.

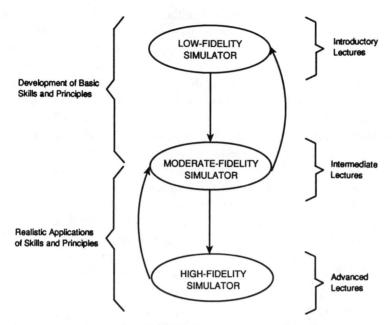

Figure 9.6. Mixed-fidelity approach to training.

The introductory lectures associated with the low-fidelity simulator serve two purposes. First, the whole concept of mixed-fidelity training is explicitly outlined for trainees. This is necessary because the approach may seem unusual to trainees, particularly because typical trainees are quite anxious to use the high-fidelity simulator. The introductory lectures also serve to explain and reinforce the basic skills and general principles that trainees are learning by experience with the low-fidelity simulator.

With the moderate-fidelity simulator, trainees begin to learn about the basic functional relationships within the equipment system of interest as well as within other similar systems. These other systems are included in order for trainees to gain an understanding of similarities among classes of equipment and how the basic skills and general principles learned earlier are applicable to more than just the equipment system of interest.

The intermediate lectures associated with the moderate-fidelity simulator also serve two purposes. First, they serve to point out and emphasize the application of basic skills and general principles to the variety of systems being simulated. Second, they begin to introduce equipment-specific information and, to a slight extent, procedures for dealing with particular events.

The high-fidelity simulator may be either a traditional simulator or a real equipment trainer. The choice depends on availability, cost, and the possibility of safely generating the events of interest using real equipment. In either case, the primary purpose of the high-fidelity simulator is to provide a totally realistic environment within which the skills and principles learned with the lower-fidelity simulators can be applied and evaluated. A secondary, but nevertheless very important, purpose of the high-fidelity simulator is to allow practice of procedures for those situations where proceduralized responses have been found to be appropriate.

As with the introductory and intermediate lectures, the advanced lectures also serve two purposes. First, procedures and other equipment-specific information are presented in more detail than in earlier lectures. The reason for this is that much of this type of information is not fully meaningful until the high-fidelity simulator is encountered. The second purpose of the advanced lectures is to explicitly contrast general strategic and tactical skills with the use of equipment-specific knowledge. In this way, operational and system knowledge are integrated and illustrated in the context of realistic problem solving.

An important aspect of the relationships among the simulators shown in Figure 9.6 is the possibility for trainees to return from a higher- to a lower-fidelity simulator. If trainees encounter situations with the high-fidelity simulator that they have great difficulty dealing with, they may be able to resolve the difficulty if they can return to the moderate-fidelity simulator to experience an approximation of the same situation. Similarly, if trainees encounter fundamental problem-solving difficulties (e.g., frequent inferential errors) when using the moderate-fidelity simulator, it may be useful for them to return to the low-fidelity simulator. Hence, the mixed-fidelity approach attempts to provide trainees with the level of fidelity necessary for learning the skills needed at each point in the training program.

Before reviewing several case studies involving application of this approach to training, it is necessary to discuss computer-based instruction. This is due to the fact that computer-based instructional environments are especially amenable to the mixed-fidelity approach.

COMPUTER-BASED INSTRUCTION

The use of computers to deliver and manage instruction emerged in the late 1960s and early 1970s (Alessi and Trollip, 1985; Rouse, 1987b). A variety of terminology has been used to denote this type of use of computers. A list of alternatives is shown in Figure 9.7. This section employs computer-based instruction (CBI) as the overall label that includes, as special cases, all of the types of uses of computers for instruction shown in Figure 9.7.

With computer-managed instruction (CMI), the computer is a learning resource manager that guides trainees' activities based on test scores and/or assessments of learning preferences. With CMI, instruction and testing are not necessarily delivered via the computer. The primary value of CMI is in scheduling efficient use of people and resources within a variety of constraints.

Computer-aided testing (CAT) focuses on assessing learning outcomes. Uses can range from simple multiple-choice tests to more sophisticated assessments of problem-solving performance. The value of CAT lies in the ease of scheduling tests, which enables avoiding artificial time constraints, as well as the possibility of more extensive testing than paper and pencil allows.

Computer-aided instruction (CAI) denotes the use of computers to present information to trainees. The form of presentation may include drill and practice, tutorials, and simulations. Trainees typically interact with the computer by responding to questions presented. Depending on the sophistication of the CAI, the course of instruction and information presented may be adapted to trainees' responses. CAI has many strengths, such as individualized and self-paced instruction, immediate feedback and potentially high trainee involvement, and low cost per trainee when the number of trainees is large.

Intelligent tutoring systems (ITS) represent an integration of artificial intelligence (AI) technology and computer-based instruction (Pstoka, Massey, and Mutter, 1988; Richardson and Polson, 1988). Typical ITS are

---

- Computer-Managed Instruction (CMI)

- Computer-Aided Testing (CAT)

- Computer-Aided Instruction (CAI)

- Intelligent Tutoring Systems (ITS)

- Embedded Training (ET)

Figure 9.7. Computer-based instruction.

composed of three models. The *expert model,* which is usually represented as an expert system, represents the goal of the training process—in other words, the trainees' behaviors are to match the experts. The *student model* represents the current state of trainees' knowledge and skills relative to the expert model. Differences between the expert and student models serve as inputs to the *instructor model,* which invokes various instructional strategies whose purpose is to remediate these differences.

Embedded training (ET) represents the integration of training within an operational system. It is usually computer-based and invoked when either time allows (i.e., slow periods operationally) or system users are unable to proceed without further instruction. There is great potential to use ET and aiding in complementary fashion, with the choice of mode (training vs. aiding) depending on several situational and personnel-related variables (Rouse, 1987a). The architecture for the design information system illustrated in Figure 8.20 depicts the use of ET via the explain and tutor functions.

SIMULATION-ORIENTED COMPUTER-BASED INSTRUCTION

CBI provides an excellent vehicle for implementing the mixed-fidelity approach to training. Simulation-oriented CBI involves embedding a simulation of the equipment system of interest within the instructional environment. The simulation provides an important "active" training component in conjunction with more "passive" instructional materials. In particular, operational knowledge can be gained by hands-on use of the simulation while system knowledge is also presented as is appropriate.

In this section, mixed-fidelity operations and maintenance training is illustrated in the context of simulation-oriented CBI. This approach to training includes low- and moderate-fidelity simulators, which are described below, and the use of a high-fidelity simulator or real equipment, depending on the application.

Low-Fidelity Simulator

In considering the essential aspects of troubleshooting equipment systems, or problem solving in complex systems in general, one specific aspect seems to be especially important. This aspect is best explained with an example. When trying to determine why component, assembly, or sub-system A is producing unacceptable results, one may note that acceptable performance of A requires that components B, C, and D perform accept-ably, since component A depends on them. Further, B may depend on E, F, G, and H, whereas C may depend on F and G, and so on. Trouble-shooting or problem solving in situations such as this example involves dealing with a hierarchy of dependencies among components in terms of their abilities to produce acceptable outputs. Abstracting the accept-able/unacceptable dichotomy with a 1/0 representation allowed the class of tasks described in this paragraph to be the basis of the low-fidelity sim-ulator.

Specifically, the low-fidelity simulator involves troubleshooting of gra-phically displayed networks. An example is shown in Figure 9.8. These networks operate as follows. Each component has a random number of inputs. Similarly, a random number of outputs emanate from each com-ponent. Components are devices that produce either a 1 or 0. All outputs emanating from a component carry the value produced by that component.

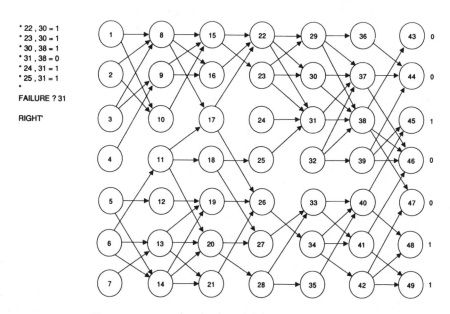

Figure 9.8. Display for low-fidelity simulator (TASK).

A component will produce a 1 if (1) *all* inputs to the component carry values of 1 and, 2) the component has not failed. If either of these two conditions is not satisfied, the component will produce a 0. If a component fails, it will produce values of 0 on all the outputs emanating from it. Any components that are reached by these outputs will in turn produce values of 0. This process continues, and the effects of a failure are thereby propagated throughout the network.

A problem begins with the computer displaying a network with the outputs indicated, as shown on the right-hand side of Figure 9.8. Based on this evidence, the trainees' task is to "test" connections between components until the failed component is found. The lower left-hand side of Figure 9.8 illustrates the manner in which connections are tested. An asterisk is displayed to indicate that trainees can choose a connection to test. Test choices are entered in the form "component 1, component 2" and result in the value carried by the connection being displayed. If trainees respond to the asterisk with a simple "return," they are asked to designate the failed component. Then, they are given feedback about the correctness of their choices. And then, the next problem is displayed.

To aid trainees in learning to use this simulator, computer aiding is available. Succinctly, the computer aid is a somewhat sophisticated bookkeeper that uses the structure of the network (i.e., its topology) and known outputs to eliminate components (i.e., by crossing them off) that cannot possibly be the fault. Also, it iteratively uses the results of tests, chosen by trainees, to further eliminate components from future consideration by crossing them off. In this way, the "active" network iteratively becomes smaller and smaller.

Owing to the fact that this simulator places total emphasis on problem structure rather than context, the acronym TASK, which stands for Troubleshooting by Application of Structural Knowledge, was chosen for it.

Moderate-Fidelity Simulator

TASK involves context-free troubleshooting in that it has no association with a particular equipment system. Moreover, trainees never see the same problem twice and, hence, cannot develop skills specific to one problem. Therefore, one must conclude that any skills that trainees develop with TASK are general, context-free skills.

However, real-life tasks are not context-free. Accordingly, the moderate-fidelity simulator is context-specific and requires trainees to use a

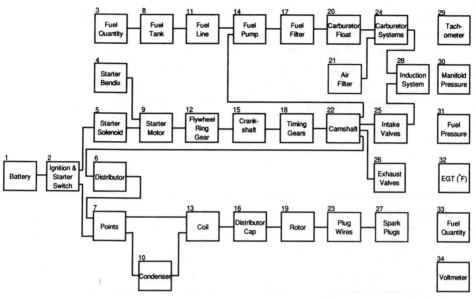

Figure 9.9. Functional block diagram for automobile engine.

functional block diagram or schematic as shown in Figure 9.9. Trainees interact with this system using the display shown in Figure 9.10. The software for generating this display is fairly general and particular systems of interest are completely specified by data files, rather than by changes in the software itself. Thus far, various automobile, aircraft, marine, communication, power, and space systems have been developed.

This simulator operates as follows. At the start of each problem, trainees are given fairly general symptoms (e.g., engine cranks but will not start). They can then gather information by checking gauges, asking for definitions of the functions of specific components, making observations (e.g., continuity checks), or by removing components from the system for bench tests. They also can replace components in an effort to make the system operational again.

Associated with each component are costs for observations, bench tests, and replacements as well as the a priori probability of failure. Trainees obtain this data by requesting information about specific components. The amount of time needed to perform observations and tests is converted to dollars and combined with replacement costs to yield a single performance measure of cost. Trainees are instructed to find failures so as to minimize total cost.

As with TASK, computer aiding is also provided with the moderate-fidelity simulator. One aspect of this aiding monitors trainees for inferential errors. When such errors occur, trainees are provided with an on-line context-specific explanation of the nature of the errors—an example of this type of feedback is shown at the bottom of Figure 9.10. Another aspect of aiding assists trainees in identifying those components in the feasible set of failures. Trainees can also ask for advice on which tests to make.

TECHNICAL SYSTEM: Car Engine		Problem 1 of 5

SYMPTOM: Cranks but Will Not Start

GAUGES	ACTIONS	COST
29 Tachometer Low	Observe (23)Plug Wires to (27)Spark Plugs Normal	$1
30 Manifold Pressure	Observe (14)Fuel Pump to (17)Fuel Filter Abnormal	$3
31 Fuel Pressure Abnormal Low	Bench Test (14)Fuel Pump Normal	$8
32 EGT (°F)	Observe (13)Coil to (16)Dist. Cap Normal	$1
33 Fuel Quantity		
34 Voltmeter Normal		

OPTIONS	
FREE	EXPENSE
INFORMATION I	OBSERVATION . . . . O
COMPARISON C	BENCH TEST. . . . . B
DESCRIPTION . . . . . . D	REPLACE. . . . . R
GAUGE. . . . . G	QUIT. . . . . Q
POSSIBLE. . . . . P	
ADVICE. . . . . A	
HELP. . . . . ?	

YOUR ACTION >   0 13-16

Error Message: It was unnecessary to check (13)Coil since you already know there is Normal Output from the (23)Plug Wires which require Normal Output from the (13)Coil

Figure 9.10. Display for moderate-fidelity simulator (FAULT).

In recognition of the generality of the software developed for the moderate-fidelity simulator, the acronym FAULT, which stands for Framework for Aiding the Understanding of Logical Troubleshooting, was chosen for the simulator.

R&D in Aviation Maintenance

The mixed-fidelity approach to training, as well as simulation-oriented CBI involving the above simulators, was developed using an aircraft power plant maintenance training program as a test bed. Over a five-year period, 10 experiments were performed involving over 300 trainees and data collection for a total of approximately 24,000 troubleshooting problems. In this section, the results of this R&D effort are summarized in terms of maintainers' troubleshooting abilities, as well as approaches to improving their troubleshooting knowledge and skills (Rouse and Hunt, 1984).

The results of this series of studies showed that maintainers are not optimal troubleshooters, although they are rational and usually systematic. In general, their deviation from optimality is related to how well they understand the problem, rather than being solely related to properties of the problem. More specifically, suboptimality appears to be due to a lack of awareness (or inability when time constraints are tight) of the full implications of available information. For example, maintainers have a great deal of difficulty utilizing information about what has *not* failed in order to reduce the size of the feasible set.

Maintenance problem solving tends to be context-dominated with familiar, or even marginally familiar, patterns of contextual cues prevailing in most problem solving. Maintainers can, however, successfully deal with unfamiliar troubleshooting situations, which is a clear indication that problem-solving skills cannot be totally context-specific. Their degree of success with unfamiliar problems depends on their abilities to transition from pattern-oriented to structure-oriented problem solving. Maintainers' abilities in the latter mode are highly related to their priorities among troubleshooting rules (i.e., which rules they apply first) rather than simply the number of rules they know.

Thus, maintainers' cognitive abilities for problem solving are definitely limited, as are the problem-solving abilities of all people. However, they have exquisite pattern recognition abilities and can cope reasonably well with ill-defined and ambiguous problem-solving situations. These abilities are very important in many real life-fault diagnosis tasks. What are needed, therefore, are methods for overcoming humans' cognitive limitations in order to be able to take advantage of humans' cognitive abilities.

Throughout this program of R&D, a variety of schemes emerged for helping humans to overcome the limitations summarized above. These schemes were evaluated both as aids during problem solving and as training methods, with evaluation occurring upon transfer to situations without aids. As noted in the earlier discussions of the low- and moderate-fidelity simulators, three types of aid were developed and evaluated.

The first type of aid was implemented with TASK and uses the structure of the network to determine the full implications of the symptoms, as well as each test, with respect to reduction of the size of the feasible set. Basically, this aid is a bookkeeper that does not utilize any information that trainees do not have—it just consistently takes full advantage of this information.

The second type of aid was also implemented within TASK. It evaluates each action by trainees, as they occur, and provides reinforcement in proportion to the degree to which the action is consistent with a context-free optimal strategy. For erroneous (i.e., nonproductive) actions, trainees receive feedback that simply notes, but does not explain, their errors. For inefficient (i.e., productive but far from optimal) actions, trainees receive feedback denoting their choices as poor or fair. Optimal or near-optimal actions yield feedback indicating choices to be good or excellent.

The third type of aid was implemented in FAULT. This aid monitors trainees' actions and checks for context-free inferential errors (i.e., errors in the sense of not using the structure of the FAULT network to infer membership in the feasible set). Although the aiding is context-free, it explains the nature of the error within the context of the problem (i.e., in terms of the structural implications of the previous actions taken) as shown on the bottom of Figure 9.10. Therefore, the feedback received by trainees not only indicates the occurrence of an error, but also includes a context-specific explanation of why an error has been detected.

The first and third types of aid can be viewed as structure-oriented bookkeeping aids, while the second type of aid is more strategy-oriented. The results of evaluating these aids for the aircraft mechanics were quite clear. The bookkeeping methods consistently improved performance, both while they were available and upon transfer to unaided problem solving. The strategy-oriented aid degraded performance and resulted in negative transfer of training, providing clear evidence of the hazards of only reinforcing optimal performance.

Several studies of transfer of training were performed. These studies involved

- Aided TASK to unaided TASK,
- Aided TASK to unaided FAULT,
- Aided FAULT to unaided TASK,
- Aided FAULT to unaided FAULT,
- Aided TASK to real equipment,
- Aided FAULT to real equipment, and
- Aided TASK plus aided FAULT to real equipment.

For the studies involving only TASK and FAULT, positive transfer of training was usually found, with the effects most pronounced for unfamiliar systems and fine-grained performance measures. Thus, the evidence is quite clear that humans can be trained to have context-free problem-solving skills that, at least partially, help them to overcome the limitations discussed earlier.

In terms of transfer of training from TASK and/or FAULT to real equipment, the results for the aircraft mechanics showed that training based on simulations such as TASK and FAULT are competitive with traditional instruction, even when traditional instruction provides explicit solution procedures for the failures to be encountered. But the issue is not really one of TASK and/or FAULT vs. traditional instruction. The important question is how these training technologies should be combined to provide a mixed-fidelity training program that capitalizes on the advantages of each technology. The remainder of this section illustrates how the mixed-fidelity approach was adapted to several applications.

Application to Supertanker Propulsion Systems

Marine engineering is a rather different domain than aircraft maintenance. One especially notable difference is the fact that shipboard engineering officers typically have to keep the propulsion system operating at the same time they attempt to detect, diagnose, and compensate for failures. In contrast, aircraft maintenance is seldom performed while the aircraft is flying.

Despite this and other differences, the mixed-fidelity approach is definitely applicable to marine engineering, especially since simulator fidelity is just as important an issue in the marine domain as in the aircraft domain (van Eekhout and Rouse, 1981). In recognition of this situation, efforts were begun to apply the mixed-fidelity approach to marine engineering training in the early 1980s (Rouse, 1982).

The initial training program was evaluated by senior marine engineers. Subsequently, the training of several hundred marine engineering officers began. In keeping with the tenets of the sales and service phase discussed in Chapter 6, I initially taught the introductory portions of the course to demonstrate to the permanent instructional staff how the mixed-fidelity approach, including TASK and FAULT, should be introduced.

The training program is conducted by Marine Safety International, Inc. and follows the basic mixed-fidelity approach outlined in Figure 9.6. Microcomputer versions of TASK and FAULT serve as the low-fidelity and moderate-fidelity simulators, respectively. A multimillion dollar full-scale engine room simulator fills the role of high-fidelity simulator.

A microcomputer is assigned to each trainee in an eight-person class. A computer-aided instruction module, also programmed on the microcomputer, helps trainees to learn about the use of the microcomputer as well as the basis of TASK and FAULT. This individualized instruction is supplemented with the aforementioned introductory lectures on the philosophy underlying mixed-fidelity training and demonstrations of TASK and FAULT.

When using FAULT, trainees diagnose failures in automobile and aircraft power plants as well as failures in marine engineering systems such as are seen in the high-fidelity simulator. By dealing with a variety of equipment systems in this way, trainees come to realize how basic skills and general principles can be applied to a wide range of problems. Evidence supporting this conclusion includes frequent remarks by trainees that strategies learned in TASK can be applied to dealing with unfamiliar systems and situations in FAULT.

This training program has been very well received by trainees over several years of ongoing delivery. This is probably due to the fact that the training program requires a high degree of involvement on the part of trainees (i.e., more doing than listening). This level of involvement is possible because of the mixed-fidelity approach where low-cost, microcomputer-based simulators are combined with a single high-cost, high-fidelity simulator. In contrast, if the training program was totally based on the high-fidelity simulator, trainee involvement would be decreased or, if class size was reduced, the throughput of the program and hence its revenues would be reduced. Thus, the mixed-fidelity approach offers potential economic advantages as well as increased training effectiveness.

Continued involvement with the evolution of this training program has led to interesting additional efforts. The TASK and FAULT training technology was also adopted by Marine Safety's parent company for use in aviation maintenance training. In addition, a series of R&D studies were performed to determine the impact of a variety of enhancements of FAULT (Su, 1985).

Application to Portable Electronic Switchboards

This application was concerned with training military personnel attending the U.S. Army Signal School at Fort Gordon (Johnson, 1987). The program of interest provides interservice training for personnel who operate and maintain telephone switchboards and related communications networks. The course of instruction within which simulation-oriented CBI was applied lasts 39 weeks, including 3 weeks on a portable electronic switchboard for use in the field. This course was selected for infusion of new training technology because the switchboard equipment represented state-of-the-art digital technology and because the annual student throughput was quite high. Further, nearly 50 percent of the course was centered on diagnostic-related tasks. Finally, real equipment was scarce and expensive, which had resulted in little hands-on experience for trainees.

This application differed from the aircraft maintenance and supertanker propulsion system applications in one primary dimension. Military trainees were not initially willing to accept or adapt to the same level of abstraction (i.e., TASK and FAULT) employed in those previous applications. Military students wanted a training device that had a high degree of physical similarity to the real equipment.

Consequently, the courseware was designed to include a range of graphics displays that realistically depicted various aspects of the switchboard. These displays, as well as associated alphanumeric information, provided the interface to FAULT. The trainees initially focused their attention on the graphically displayed information. However, it soon became apparent to them that FAULT provided the most information and feedback—eventually, therefore, FAULT received considerable use.

Evaluation of the courseware at Fort Gordon showed that it was a reasonable substitute for a portion of the real-equipment training and practice. For several years, this courseware has been an approved portion of the official program of instruction.

Application to Emergency Diesel Generators

This application began with an extensive survey of trainers from nuclear power plants (Johnson et al., 1985). Results showed that there was a need for improved diagnostic-related training for both operations and maintenance personnel. Emergency diesel generators were chosen as the equipment system for which simulation-oriented CBI would be developed and evaluated.

Based on the experiences with the switchboard courseware discussed above, it was decided to emphasize computer-generated graphics to deliver trainees to FAULT-based diagnostic training. Consequently, a large portion of the courseware was devoted to color graphic representations of control panels, equipment, and facility spaces. Therefore, beyond FAULT, the courseware included a considerable amount of computer-aided instruction.

The courseware was extensively evaluated, both during the development process and subsequently. This evaluation is described in detail in a later section of this chapter. It is useful to note two results, however. First, while trainees found the color graphics very appealing, they eventually spent almost three-quarters of their time using FAULT—this is totally consistent with the findings for the switchboard courseware. Second, the results of the evaluation were sufficiently impressive (discussed later) to lead several nuclear utilities to adopt this training program, including Duke Power Company where the evaluations were conducted.

Application to Space Shuttle Fuel Cell

The most recent application of simulation-oriented CBI was to operations and maintenance training for the space shuttle fuel cell (Johnson et al., 1988). This application involved using FAULT, with the addition of appropriate rule-based knowledge, as the basis for an intelligent tutoring system as defined earlier in this chapter. The resulting system, though relatively rudimentary, provided convincing evidence that the basic functionality of an intelligent tutoring system could be provided in a timely manner and at quite reasonable cost. This system is currently being evaluated.

To move beyond a rudimentary intelligent tutoring system for equipment operation, several enhancements are needed. A dynamic representation of the equipment system that allows trainees to exercise control is important. Further, the instructor model has to be extended to provide instruction that is more sensitive to the specific nature of trainees' actions and errors. A good example of such an ITS is Fath's system for training troubleshooting of marine power plants (Fath, Mitchell, and Govindaraj, 1987).

EVALUATION OF TRAINING

The emergency diesel generator courseware was evaluated using the methodology described in Chapter 7. This included evaluation of compatibility, understandability, and effectiveness. This section describes the evaluation procedures and the results (Maddox, Johnson, and Frey, 1986).

Compatibility

The evaluation of compatibility was concerned with the issues listed in Figure 9.11. This evaluation was performed during development using the guidelines of Frey and his colleagues (Frey et al., 1984). A variety of minor problems—concerned with, for example, display write times and ambient lighting—were identified and corrected.

Understandability

The understandability evaluation focused on the issues in Figure 9.12. These issues were pursued in a two-part evaluation. First, the structure,

• Character Size	• Color contrast
• Character Density	• Keyboard Location
• Symbol Size	• Time to Write Display
• Refresh Rate	• Response Time
• Contrast	• Ambient Lighting
• Color Usage	• Glare

Figure 9.11. Compatibility issues.

```
• Courseware Structure
• Student Interaction
• Word and Abbreviation Usage
• Color Usage
• Labels and Part Names
• Feedback
```

Figure 9.12. Understandability issues.

format, and content of the human–computer interaction were analyzed during development. Second, trainees' use of the courseware was evaluated at the training center, where formal evaluations were conducted.

Two types of problem were identified. One concerned the fact that FAULT required interaction via a keyboard, whereas the remainder of the courseware could be accessed with a mouse. This created some initial problems for the trainees, but they quickly learned to deal with the two entry devices.

The second type of problem concerned labeling of equipment components, as well as the names of commands in FAULT. Inappropriate labels and ambiguous command names were modified as necessary. All compatibility and understandability problems were corrected prior to evaluation of effectiveness.

Effectiveness

The evaluation of effectiveness concerned the issues listed in Figure 9.13. This evaluation involved a comparison of the simulation-oriented CBI to instruction that was much more traditional. Immediately subsequent to training with one of these methods, trainees performed five troubleshooting problems with the real equipment.

For safety reasons, trainees actually only walked next to the diesel generator and told the evaluator what they would do. The evaluator, who was not aware of how they had been trained, recorded their decisions and verbally informed them of what the results of their actions would have been. Trainees had a maximum of 20 minutes to locate each of the five hypothetical failures.

```
• Immediate and Retention
• Knowledge
• Skills
• Attitude
• Job Performance
• License Exam
```

Figure 9.13. Effectiveness issues.

An evaluation of training retention was conducted 20 weeks after the immediate transfer of training evaluation. Each trainee again performed five troubleshooting problems, two of which were identical to the earlier problems. As before, trainees' troubleshooting involved hypothetical problems with the diesel generator.

Trainees' performance on these troubleshooting problems was analyzed using several measures, including solution time, number of actions, number of errors, an overall performance index, and an evaluator's performance rating. Analysis of the transfer of training data collected immediately after training indicated no statistically significant differences between the simulation-oriented CBI and traditional instruction for these overall performance measures. However, as noted in Chapter 7, a fine-grained analysis indicated that maintainers who trained with traditional instruction committed serious errors at a rate five times greater than the other three groups—the other three groups included maintainers trained with CBI and operators trained with either method. Thus, the CBI had a bigger impact on maintainers than operators.

As noted above, the data on retention of training were collected 20 weeks later. Analysis of this data indicated that personnel trained with CBI solved the real equipment troubleshooting problems 23 percent faster than those trained with traditional instruction. The maintainers who trained with traditional instruction still committed serious errors at a rate five times greater than the other groups.

The results of this evaluation effort are important for several reasons. First, the advantages of the simulation-oriented CBI, in terms of reduced diagnostic time and fewer serious errors, provided convincing evidence for nuclear utilities to adopt this courseware. It also caused them to view simulation-oriented CBI more favorably in general. Second, this study provided seldom-collected evidence of the positive longer-term effects of CBI—this possibility is frequently noted, but data to support the assertion are infrequently gathered. Third, this study lent further support for the guidelines and principles discussed in Chapter 7 (Figs. 7.4 and 7.5)—quite simply, it is often very difficult to demonstrate the impact of features of products and systems by only considering global or overall measures.

TRAINING VS. AIDING

From the discussion thus far in this chapter, it should be clear that there are a variety of ways to train people. From the discussion in Chapter 8, it should be equally clear that there are a variety of ways to aid people. In general, more highly trained people need less aiding, and those with less training require more aiding. The resulting trade-off is obvious. How should one balance training and aiding to accomplish the operational objectives of the system in a cost-effective manner? This section discusses how this question can be answered.

Training vs. aiding trade-offs are usually pursued in terms of performance, time, and cost. If a particular combination of training and aiding provides a performance advantage compared to alternative mixes, it will be chosen if time and costs are not adversely affected. Quite frequently, performance requirements are fixed. The primary issues then concern enabling personnel to achieve performance objectives as quickly and inexpensively as possible.

Within such analyses, the central contrast between training and aiding is quite clear. Training has the advantage that it can produce more flexible personnel who can cope with changing demands. The disadvantages of training are its recurring costs.

Aiding has the advantage that, with the exception of maintaining the aid, its recurring costs are relatively low. It may, however, have higher capital costs of acquisition. Aiding is likely to have the additional disadvantage of being fairly rigid relative to new demands. However, intelligent systems technology, such as discussed in Chapter 8, has the potential to broaden the capability and adaptability of traditional aids.

An interesting interaction between training and aiding concerns the training requirements imposed by aiding. For other than very simple types of aiding, personnel must be trained to use the aiding appropriately. These training requirements can be particularly subtle for tasks where the applicability of the aiding varies with situations. The intelligent cockpit is a good example of such a system—we are about to initiate a project to determine how pilot training requirements should change once the intelligent cockpit is fielded.

A variety of factors affect whether training or aiding, or more frequently some combination, is the most appropriate way to achieve performance objectives (Rouse, 1985, 1987a). These factors include characteristics of people, equipment, tasks, organizations, and the operating environment. Several R&D efforts have focused on developing methods of trading off the advantages and disadvantages of training and aiding as a function of these characteristics as well as life-cycle costs.

One of the earliest systematic approaches was the development of structured guidelines for mapping from the above characteristics to training and aiding alternatives, including combinations of training and aiding (Booher, 1978a, 1978b; Foley, 1978). These guidelines take the form of flowcharts that proceduralize trade-off decisions, but unfortunately do not allow fine-grained trade-offs. For example, significant but nonmajor changes in system complexity via redesign would be unlikely to change the recommendations provided by these guidelines.

Somewhat more recently, the comparability approach was developed to enable analyzing the design of a new system in terms of deviations from an existing system (Weddle, 1986). This approach assumes that an existing baseline system, including the training and aiding associated with it, provides a good starting point. This assumption is probably acceptable for modest updates of existing systems or changes involving well-understood technology.

Alternative Approaches

An in-depth review was performed of the range of possible approaches to formulating and resolving training vs. aiding trade-offs (Rouse and Johnson, 1989). It was concluded that realistically complex training vs. aiding trade-offs cannot be addressed with a single method. Instead, it is more appropriate to consider how approaches can be integrated. Three integrated or composite approaches were developed.

The composite approach depicted in Figure 9.14 provides a rough approximation of the ways in which training vs. aiding trade-offs are currently pursued. Manpower, personnel, and training (MPT) data bases

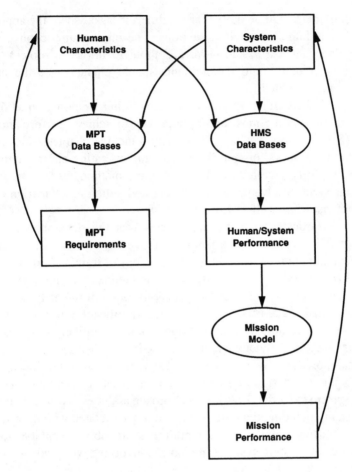

Figure 9.14. Status quo analysis.

denote both data bases and a variety of spreadsheetlike tools associated with these data bases. Human–machine system (HMS) data bases denote both formal and informal compilations of past experiences and experiments. Mission models typically involve computer simulations of the interactions of many people and equipment systems, as well as some representation of the operational environment. The types of models discussed in Chapter 5 potentially could provide inputs to mission models—for example, human performance parameters.

An especially interesting aspect of Figure 9.14 is the *relative* independence of the feedback loops. If mission performance is unacceptable, system characteristics will tend to be modified independent of the impact on MPT requirements. Similarly, if MPT requirements are excessive, human characteristics will tend to be modified (perhaps via selection and classification criteria) without *direct* knowledge of the impact on mission performance.

As a result of the structure of Figure 9.14, MPT analysts and HMS designers have relatively little in common. They do not share any methods, tools, or models. Moreover, they often come from different disciplines (i.e., psychology and engineering) and consequently employ different concepts, jargon, and so on. From this perspective, it is not surprising that systems emerge with latent MPT problems that lead to poor mission performance and/or costly redesign.

Figure 9.15 employs a computational human–machine system model to integrate the MPT/HMS design process. In the training vs. aiding context, this model or, more likely, set of models is concerned with human–system performance with various levels of aiding and alternative approaches to training. The models are used to predict the impacts of different combinations of aiding and training.

With a performance-based analysis, MPT analysts and HMS designers have to communicate in order to proceed. In a sense, the HMS model(s) provides a unifying metaphor which both types of individuals utilize and influence in somewhat different ways. Beyond facilitating communication, this approach can be the basis for providing powerful computational support for pursuing trade-offs.

A significant limitation of the performance-based approach to analysis is the possible need to go beyond performance predictions and study the behaviors underlying performance measures. This can be particularly important when the impact of training is of concern.

As noted at the beginning of this chapter, training can be defined as a process of managing people's experiences so that they gain the knowledge and skills that give them the potential to perform. If one is concerned with the extent to which a given approach to training results in acquisition of the requisite knowledge and skills, then it may be necessary to examine the process of learning as it is affected by the training regime of interest. This potentially can be accomplished by using a computational model that, via simulation, experiences the training and acquires knowledge and skills. While the state of the art is such that this approach is not yet feasible for most complex tasks, progress in computational modeling is sure to provide such formulations in the foreseeable future (Baron and Kruser, 1990;

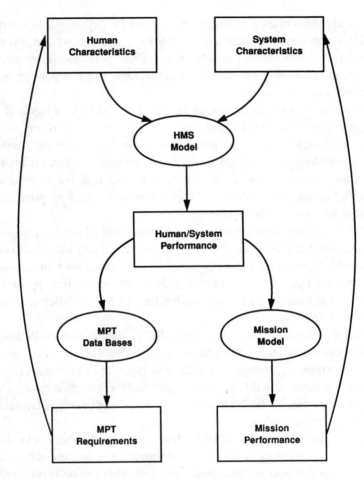

Figure 9.15. Performance-based analysis.

Elkind et al., 1989; McMillan et al., 1989; Rouse and Johnson, 1989).

The potential use of behavioral simulations is illustrated in Figure 9.16. Rather than predicting performance, in this approach behavior is predicted and performance is calculated. Use of behavioral simulations leads to a variety of issues such as representativeness of scenarios, number of runs, appropriate statistics, and so forth. Of course, these issues are not new—they are similar to the issues involved in using human-in-the-loop simulators for experimental studies such as discussed in Chapter 7.

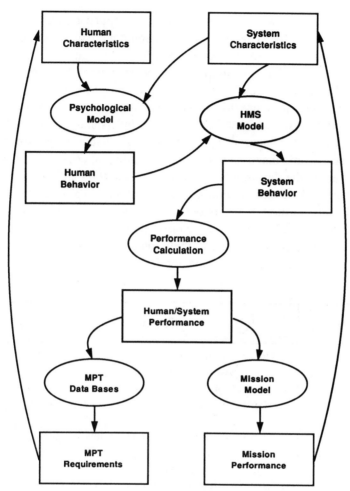

Figure 9.16. Behavior/performance-based analysis.

An Analysis Environment

The types of analysis depicted in Figures 9.14 through 9.16 require managing many issues, methods, and evolving predictions of the impacts of alternative trade-offs. Furthermore, these approaches to trade-off analysis typically require access and utilization of a wealth of information. How could an analyst be reasonably expected to pursue such a comprehensive approach?

The obvious answer to this question is a computer-based version of the analysis framework. This analysis "environment" supports the types of methods and tools depicted in Figures 9.14 through 9.16. Structure is provided to the analysis environment by providing and supporting the trade-off analysis process summarized in Figure 9.17.

The 15 steps in Figure 9.17 can be explained in the context of much of the material presented in earlier chapters. Understanding the job (step 1) is clearly part of the naturalist phase of design (Chapter 3). Task decomposition (step 2) and mapping to training and aiding alternatives (steps 4 and 5) draw upon the discussions in Chapters 8 and 9 on aiding and training, respectively. Assessing limitations and abilities (step 3) includes elements of Chapters 3 through 5, where one iteratively comes to understand users. Steps 8 through 12 deal with measures, models, and computations. Analytical approaches to these steps are discussed in Chapter 5, and empirical approaches are considered in Chapter 7.

STEP	ACTIVITIES
1	Understand the Job
2	Decompose via Task Taxonomy
3	Assess Limitations, Abilities, and Preferences
4	Map Limitations, Etc., to Taxonomy of Training Alternatives
5	Map Limitations, Etc., to Taxonomy of Aiding Alternatives
6	Make Obvious Choices
7	Coalesce Trade-offs Involving Interdependent Tasks
8	Choose Measures of Performance
9	Choose Input–Output Representations
10	Identify Requisite Structure and Parameters for Representations
11	If Necessary, Represent Learning Process
12	Apply Methods of Analysis to Representations
13	Interpret Results
14	Compile Assumptions and Consequences of Trade-offs
15	Form Sets of Trade-offs with Consistent Assumptions and Consequences Regarding Personnel

Figure 9.17. Trade-off analysis process.

This leaves steps 6, 7, 13, 14, and 15. While steps 14 and 15 are primarily bookkeeping tasks, steps 6, 7, and 13 require a reasonable level of expertise, either on the part of the user or embedded in the system. The latter possibility is considered in the following discussion.

The process in Figure 9.17 is viewed as a *nominal* sequence of steps that will enable an analyst to formulate and resolve training vs. aiding trade-offs. The term nominal is italicized to emphasize that this 15-step process is not necessarily a lockstep sequence. Depending on an analyst's level of experience, significant sequential variations are likely, as well as variations in the ways the analysis environment supports users.

Types of User. The flexibility of the analysis process, and the ways in which this process is supported, are essential for successfully assisting the range of users envisioned for the analysis environment. Three types or classes of user are envisioned: novice, journeyman, and expert.

Successful formulation and resolution of training vs. aiding trade-offs require five types of expertise:

- Domain—characteristics of jobs/tasks, equipment, and so on.
- Training—organizational practices, training methods and technologies, and so forth.
- Aiding—organizational practices, aiding methods and technologies, and so on.
- Discipline—psychology, engineering, technical education, and so forth.
- Methodology—analysis and modeling methods, analysis process in Figure 9.17.

In any of these five areas where a specific user is lacking expertise, the analysis environment must augment the user's abilities and performance.

Obviously, it is improbable that any particular user of the analysis environment will be strictly a novice, journeyman, or expert in *all* five of these areas. Consequently, the analysis environment must be capable of supporting a particular user in different ways depending on the match of that user's expertise to the step of the analysis process being pursued at that point in time.

It is essential to recognize that all three types of users are unlikely to be responsible for the full range of possible training vs. aiding trade-offs. This

	PHASE OF SYSTEM LIFE-CYCLE	
CONSEQUENCES OF TRADE-OFF	**UPSTREAM DESIGN**	**DOWNSTREAM MPT**
HIGH (e.g., Aircraft Pilots)	Experts	Journeymen
LOW (e.g., Security Guards)	Journeymen	Novices

Figure 9.18. Factors influencing type of users.

notion is captured in Figure 9.18 in terms of two attributes: phase of system life-cycle and consequences of trade-off. Based on this construct, one can conclude that it is extremely unlikely, for example, that novices will be responsible for determining the nature of pilot training. In particular, novices will not, in this example, be responsible for deciding whether pilot training should be enhanced or the aircraft computers should be more sophisticated.

Figure 9.18 also provides a basis for concluding that novices will not be responsible for trade-offs during early phases of system design, when the system definition tends to be both conceptual and fluid. In contrast, once a system has been fielded and MPT problems have arisen, it is likely that novices may be responsible for formulating and resolving trade-offs among alternative ways of dealing with these problems. It is important to note that the word *system* is being used in a broad sense to include equipment and, for example, organizational systems.

The constraints embodied in Figure 9.18 are important in that they imply that the analysis environment need not be capable of supporting analysts to perform very complex analyses for which they have little if any background. This assumption is crucial to avoiding immense theoretical and practical problems, such as embedding sufficient knowledge in the analysis environment to allow completely naive users to solve very sophisticated problems.

Nature of Support. Figure 9.19 illustrates the types of support that the analysis environment provides to each type of user for each step of the process (Rouse et al., 1989). The entries in this tabulation are meant to be representative rather than comprehensive. There are several especially significant differences among the types of support provided in the three columns of Figure 9.19:

- Novice
  - — Environment initiates interactions
  - — User fills forms or templates
  - — Environment provides single recommendation
- Journeyman
  - — Mixed environment/user initiative
  - — User chooses among and fills in alternative forms or templates
  - — Environment provides multiple recommendations
- Expert
  - — User initiates interaction
  - — User creates forms or templates
  - — Environment evaluates user's alternatives.

Beyond the above differences, there is a subtle and very important issue underlying user–environment interaction. It is obvious from Figure 9.19 that the nature of support varies with the step being supported (e.g., providing taxonomies vs. exercising models). Less obvious is the aforementioned notion that a particular user's level of expertise will not be uniform across steps. Consequently, users must be allowed and supported when moving across rows on Figure 9.19, not just up and down columns. In other words, an expert, for example, must be able to revert to a journeyman or novice for a specific step in the process, and be supported accordingly.

Current Status of Trade-off Analysis Tool

The naturalist and marketing phases for the trade-off analysis tool were discussed in Chapters 3 and 4, respectively. This section has presented the concept for this analysis tool as of this writing. Several aspects of the evolution of this concept are of note. First, our evolving understanding of the nature of training vs. aiding trade-offs resulted in the concept of a "tool" being replaced by an "environment" that provides access to many tools, methods, and data bases. Second, the requirement to be able to

USER STEP	NOVICE	JOURNEYMAN	EXPERT
1	Standard Job Descriptions Supplied	Job Description and Templates Compared	Job Description Templates Supplied
2	Task Descriptions Supplied and Used	Descriptions and Taxonomies Compared	Alternative Task Taxonomies Supplied
3	User Supplies Minimal Inputs	Relevant Limitations, Abilities, Etc., Recommended	Choice Among Task Taxonomies Maps to Taxonomy of Limitations, Abilities, Etc.
4	Direct Mapping to Training Alternatives	Alternative Mappings to Training Taxonomy Recommended	Choices Among Limitations, Etc., Maps to Training Taxonomy
5	Direct Mapping to Aiding Alternatives	Alternative Mappings to Aiding Taxonomy Recommended	Choices Among Limitations, Etc., Maps to Aiding Taxonomy
6	Choice(s) Recommended	Alternative Choices Recommended	User's Choice(s) Evaluated
7	Standard Formulation Supplied	Alternative Formulations Recommended	User's Formulation Evaluated
8	Standard Performance Measures Supplied	Standard Measures and Taxonomy Compared	Taxonomy of Relevant Measures Supplied
9	Preconfigured Models Supplied for Standard Formulations and Measures	User Supplies Criteria and Rank-Ordered Set of Alternative Models Supplied	User Chooses Among Set of Relevant Modeling Approaches
10	User Supplies Parameters for Limited Number of Equipment and Personnel Characteristics	User Chooses Among Alternative Templates and Makes Straightforward Modifications	User Configures Structure of Model and Obtains Parameter Estimates
11	Standard Learning Curves Used at Most	User Chooses Among Learning Curve Models and Behavior/Learning Models	User Configures Behavior/Learning Model and Obtains Knowledge and Skills
12	Preconfigured Models Exercised	Models Exercised with Options for Range of Outputs and Intermediate Results	Models Exercised With Wide Range of Options Including Debugging
13	Predictions Compared to Performance Requirements	Prediction Compared to Typical Model Outputs (for Verification) and Performance Requirements	Facilities for Sensitivity Analyses and Optimization Supplied
14	Audit Trail Compiled and Available	Audit Trail Presented and User Edits and Augments	Audit Trail Presented and User Assesses Consistency and Acceptability of Analyses
15	Sets of Trade-offs Recommended	Alternative Sets of Trade-offs Recommended	User's Set of Trade-offs Evaluated

Figure 9.19. Nature of support.

support a wide range of users led to a need to provide a strong advisory component within the environment, especially for novices. This need is the driving force behind current efforts on the project to develop and embed appropriate knowledge bases in the analysis environment so that the advisory functions within the upcoming third prototype will have sufficient breadth and depth. Thus, this project is in the conceptual design step of the engineering phase, with some marketing measurements remaining once the next prototype is available.

SUMMARY

Training and aiding are leverage points. Appropriate use of this leverage can make the difference between an adequate product or system and an innovative solution to customers' and users' problems. While either training or aiding can provide this leverage, the greatest gains are achieved by providing a suitable balance. This balance should weigh the limitations, abilities, and preferences of users vs. operational requirements, resource constraints, and technology available. The human-centered concepts and methods discussed in this chapter, as well as other chapters, provide the means for achieving this balance.

One primary issue remains in our pursuit of design success. This issue is technology transition. What are the avenues and barriers to transitioning technology from R&D to design, and from design to the marketplace? To answer this question, the roles of humans in design success must be considered from a broader perspective.

REFERENCES

Alessi, S. M., and Trollip, S. R. (1985). *Computer-based instruction: Methods and development.* Englewood Cliffs, NJ: Prentice-Hall.

Baron, S., and Kruser, D. S. (Eds.), (1990). *Human performance modeling.* Washington, DC: National Academy Press.

Booher, H. R. (1978a). *Job performance aids: Research and technology state-of-the-art* (Rept. TR-78-26). San Diego: Navy Personnel Research and Development Center.

Booher, H. R. (1978b). *Job performance aid selection algorithm: Development and application* (Rept. TN-79-1). San Diego: Navy Personnel Research and Development Center.

Booher, H. R. (Ed.) (1990). *MANPRINT: An approach to systems integration.* New York: Van Nostrand Reinhold.

Booz-Allen and Hamilton (1985). *Air Force manpower, personnel, and training (MPT) systems model course.* Wright-Patterson Air Force Base, OH: Deputy for Simulators.

Elkind, J. W., Card, S. K., Hochberg, J., and Huey, B. M. (Eds.) (1989). *Human performance models for computer-aided engineering.* Washington, DC: National Academy Press.

Fath, J. L., Mitchell, C. M., and Govindaraj, T. (1987). Evaluation of an architecture for ICAI programs for troubleshooting in complex dynamic systems. *Proceedings of the 1987 IEEE International Conference of Systems, Man, and Cybernetics,* pp. 1145–1149.

Fath, J. L., and Rouse, W. B. (1985). An approach to training for operation and maintenance in large-scale dynamic environments. *Proceedings of the 1985 IEEE International Conference on Systems, Man, and Cybernetics,* pp. 532–536.

Foley, J. P. (1978). *Executive summary concerning the impact of advanced maintenance data and task oriented training technologies on maintenance, personnel, and training systems* (Rept. TR-78-24). Wright-Patterson Air Force Base, OH: Air Force Human Resources Laboratory.

Frey, P. R., Sides, W. H., Hunt, R. M., and Rouse, W. B. (1984). *Computer-generated display system guidelines.* Volume 1: *Display design.* (Rept. NP-3701, Vol. 1). Palo Alto, CA: Electric Power Research Institute.

Gagne, R. M., Briggs, L. J., and Wager, W. W. (1988). *Principles of instructional design.* New York: Holt, Rinehart, & Winston.

Johnson, W. B. (1987). Development and evaluation of simulation-oriented computer-based instruction for diagnostic training. In W. B. Rouse (Ed.), *Advances in man-machine systems research* (Vol. 3, pp. 99–127). Greenwich, CT: JAI Press.

Johnson, W. B., Maddox, M. E., Rouse, W. B., and Kiel, G. C. (1985). *Diagnostic training for nuclear plant personnel.* Volume 1: *Courseware development* (Rept. NP-3829, Vol. 1). Palo Alto, CA: Electric Power Research Institute.

Johnson, W. B., Norton, J. E., Duncan, P. C., and Hunt, R. M. (1988). *Development and demonstration of Microcomputer Intelligence for Technical Training (MITT)* (Rept. TP-88-8). Brooks Air Force Base, TX: Air Force Human Resources Laboratory.

Maddox, M. E., Johnson, W. B., and Frey, P. R. (1986). *Diagnostic training for nuclear plant personnel.* Volume 2: *Implementation and evaluation* (Rept. NP-3829, Vol. 2). Palo Alto, CA: Electric Power Research Institute.

McMillan, G. R., Beevis, D., Salas, E., Strub, M. H., Sutton, R., and van Breda, L. (Eds.) (1989). *Applications of human performance models to system design.* New York: Plenum.

Morris, N. M., and Rouse, W. B. (1985a). The effects of type of knowledge upon human problem solving performance in a process control task. *IEEE Transactions on Systems, Man, and Cybernetics, SMC-15,* 698–707.

Morris, N. M., and Rouse, W. B. (1985b). Review and evaluation of empirical research in troubleshooting. *Human Factors, 27,* 503–530.

Pstoka, J., Massey, L. P., and Mutter, S. A. (Eds.) (1988). *Intelligent tutoring systems: Lessons learned.* Hillsdale, NJ: Lawrence Erlbaum Associates, Inc.

Richardson, J. R., and Polson, M. C. (Eds.) (1988). *Foundations of intelligent tutoring systems.* Hillsdale, NJ: Lawrence Erlbaum Associates, Inc.

Rouse, W. B. (1982). A mixed-fidelity approach to technical training. *Journal of Educational Technology Systems, 11,* 103–115.

Rouse, W. B. (1985). Optimal allocation of system development resources to reduce and/or tolerate human error. *IEEE Transactions of Systems, Man, and Cybernetics, SMC-15,* 620–630.

Rouse, W. B. (1987a). Model-based evaluation of an integrated support system concept. *Large-Scale Systems, 13,* 33–42.

Rouse, W. B. (Ed.) (1987b). *Advances in man-machine systems research* (Vol. 3: *Training systems*). Greenwich, CT: JAI Press.

Rouse, W. B., Frey, P. R., Wiederholt, B. J., and Zenyuh, J. P. (1989). *The TRAIDOFF concept.* Norcross, GA: Search Technology, Inc.

Rouse, W. B., and Hunt, R. M. (1984). Human problem solving in fault diagnosis tasks. In W. B. Rouse (Ed.), *Advances in man-machine systems research* (Vol. 1, pp. 195–222). Greenwich, CT: JAI Press.

Rouse, W. B., and Johnson, W. B. (1989). *Computational approaches for analyzing trade-offs between training and aiding.* Brooks Air Force Base, TX: Air Force Human Resources Laboratory.

Su, Y-L. (1984). *A review of the literature on training simulators: Transfer of training and simulator fidelity* (Rept. 84-1). Atlanta: Center for Man-Machine Systems Research, Georgia Institute of Technology.

Su, Y-L. (1985). *Modeling fault diagnosis performance on a marine power plant simulator* (Rept. 85-3). Atlanta: Center for Man-Machine Systems Research, Georgia Institue of Technology.

van Eekhout, J. M., and Rouse, W. B. (1981). Human errors in detection, diagnosis, and compensation for failures in the engine control room of a supertanker. *IEEE Transactions on Systems, Man, and Cybernetics, SMC-11,* 813–816.

Weddle, P. D. (1986). Applied methods for human-systems integration. In J. Zeidner (Ed.), *Human productivity enhancement* (pp. 332–363). New York: Praeger.

Chapter **10**

Technology Transition

In the preceding nine chapters, this book has described and illustrated seven measurement issues whose appropriate resolution establishes the basis for designing successful products and systems. The four-phase framework, and associated concepts, principles, and guidelines, supply the means for appropriate resolution of these seven issues. The leverage points of evaluation, aiding, and training provide opportunities for high value-added innovations.

Adoption of the human-centered philosophy and use of the methods presented in this book will satisfy the *necessary* conditions for design success. However, this commitment may not be *sufficient for success.* This possibility can be explained in terms of the differences in the stakeholders in the outcome of design—the designed *product*—and the stakeholders in the *process* of design.

Satisfying the necessary conditions outlined above will assure that the product or system meets the needs of users, customers, and other stake-holders in the outcome of design. But designers, managers, investors, and other stakeholders in the process of design have additional concerns. One concern of special importance is how technology can best be infused into the design process to assure that *inventions,* that is, great new ideas, succeed in becoming *innovations,* that is, great new products and systems.

This concern can be viewed as primarily a management issue. How can the overall design process be planned and managed so that invention is infused in the process at appropriate places and times, and thereby assures transitions to successful innovations? This question can be approached on several levels, ranging from corporate strategic planning to functional engineering management. These topics are beyond the scope of this book, with one important exception.

This topic concerns managing the transition of technology in Figure 10.1, which appeared earlier as Figure 2.8 in Chapter 2. There are several aspects of this transition process. First, how can new technology from the laboratory be transitioned into the process? In other words, how can R&D results make it to the "technology feasibility" stage depicted in Figure 10.1?

Second, how can the transitions from feasibility to development to refinement be managed? To an extent, the process depicted in Figure 10.1 and detailed throughout this book provides the basis for this aspect of managing transitions. However, it is also necessary that this process dovetail with the first aspect described above, and the following third aspect of managing transitions.

This third aspect concerns assuring that technology that has progressed through feasibility, development, and refinement results in innovative

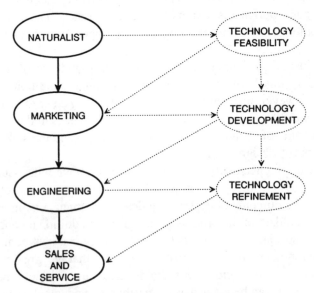

Figure 10.1. A framework for measurement.

products and systems in the marketplace. In other words, successful technology transition involves successful transitions from R&D to design *and* successful transitions through design into the marketplace.

This chapter addresses the technology transition process. The overall issues are first described, with particular emphasis on the behavioral and organizational phenomena underlying these issues. This discussion affords a foundation for suggesting how barriers to innovation can be lowered and overcome.

TRANSITIONS FROM R&D TO DESIGN

When one first compares R&D to design, the differences seem amazing and perhaps unsettling. Researchers seem like unrealistic purists, while designers appear to be unmitigated pragmatists. At the risk of oversimplifying, researchers want to "do it right" or not at all, and designers will often try anything that promises to "keep things going" in the right direction.

In fact, this is indeed a simplistic explanation of the differences between research and design. A more balanced view can be obtained by comparing research and design in terms of goals, approaches, and criteria (Rouse, 1985). The multidimensional comparison that follows provides the basis for subsequent detailed consideration of competing perspectives of system design and evaluation.

Goals: Understanding vs. Improvement

Research and design have different goals. The goal of research is *understanding* in the sense of contributing facts, principles, and laws to the general knowledge base. This knowledge should be broadly applicable and not limited in relevance to specific situations.

In contrast, design is primarily concerned with incremental *improvement* of particular situations, often to gain or maintain a market advantage. Not only are these situations specific, they are also "real," in contrast to the many "hypothetical" situations envisioned in basic research. While the goal of incremental improvement may involve accessing the general knowledge base, seldom is there any concern with augmenting this knowledge base. Solution of the specific problem at hand dominates all other considerations.

Of course, understanding and improvement are not inconsistent goals. Thus, it would seem that symbiosis is possible. However, not infrequently basic research in pursuit of understanding leads to formulations or theories that describe everything in general but nothing in particular. On the other hand, any potential generality of efforts aimed at practical improvements tends to be overwhelmed by the contingencies of the specific situation of interest. As a consequence, any general formulation (appropriate or otherwise) will seem to be of marginal value to the designer who has focused on the idiosyncrasies of the current problem. It is quite possible that this focusing may be what limits practical improvements to being incremental in nature.

Therefore, the goals of research lead to an emphasis on identifying common attributes of broad classes of problems, whereas the goals of design cause a focusing on the unique attributes of each problem. This key difference would seem to provide ample opportunities for a lack of communication between research and design.

Approaches: Scaling Down vs. Scaling Up

Given different goals, it is not surprising that researchers and designers approach their pursuits rather differently. In research, the tendency is to study *problems that are scaled down* to fit the methods being investigated. Simplifying assumptions are made in order to allow rigorous formulations. Problem-specific contingencies are eliminated in order to promote experimental control. Generality, relative to broad classes of problems, is achieved by avoiding anything specific to particular contexts.

In contrast, designers' pragmatic orientation toward improvement often results in the use of *methods that are scaled up* to fit the problem of interest. In other words, in order to approach the problem at hand, it may be necessary to utilize methods beyond their proven ranges of applicability. This approach is deemed to be justified if the desired improvement results.

Considering researchers' orientation toward general methods and designers' orientation toward specific problems, the above difference is quite natural. However, it has the unfortunate consequence that most research results are not in a form that is readily useful to designers. The scaling down of problems to fit methods often produces a myopic view of complex phenomena as simply being collections of independent elementary phenomena. In reality, the interdependencies of phenomena tend to be the essence of complexity. Accordingly, for example, when one combines all of the research findings on human performance in display reading and control manipulation, one does not get the feeling that the essence of aircraft piloting has been captured. Consequently, designers, whose orientation does not involve scaling down the problem, do not see the basic research results as relevant.

Criteria: Individual vs. Team

Beyond differences in goals and approaches to achieving goals, research and design differ in the criteria employed for judging the success of efforts. The research community emphasizes *individual accomplishment.* Reports, articles, books, and citations are counted. Quantitative and relatively objective, but not necessarily appropriate, criteria are used for decisions about salaries, promotions, and awards.

Success in design is more concerned with *organizational teamwork.* Evaluation is based on having managed or otherwise contributed to projects being completed, products being successful, and profits being realized. Qualitative and relatively subjective criteria are used to relate such accomplishments to subsequent increases of responsibilities and promotions.

The above differences in criteria can be problematic for cooperation between researchers and designers. The researcher's career is likely to depend on clearly documented individual accomplishments. Participation in a design effort seldom provides such documentation. Similarly, generalizing of results and publishing an article is often of marginal value to the designer.

While the research and design communities both have internally consistent values (i.e., goals, approaches, and criteria), they both are, in different ways, rather shortsighted. They both are working on similar problems from different perspectives. Research is attempting to understand problems from a technical point of view, whereas design is trying to solve problems incrementally within an organizational framework. The two perspectives should meet at some point in the design and evaluation process.

Design: Opportunity vs. Requirement

Design is the process of synthesizing a solution to a problem. Considering the contrasting perspectives discussed above, it is fairly clear that researchers tend to look at this problem-solving activity as an *opportunity for understanding,* while designers will view it as a requirement for improvement.

This distinction has important implications. Given an opportunity for understanding, researchers will think of "what might be" and try to *optimize.* In contrast, the requirement for improvement will cause designers to focus on "what is" and attempt to "satisfice" in terms of an incremental improvement.

In order to explore "what might be," researchers will be *objectives-driven* and, as emphasized earlier, tend to scale down the problem in order to make optimization feasible. On the other hand, designers' focusing on "what is" will lead to *constraints-driven* design that, as explained above, often requires scaling up of methods to satisfy the constraints. Hence, while the primary concern of researchers will be the assurance of the "best" solution—even if the problem statement must be compromised—designers' concerns will typically be dominated by budgets, credibility (e.g., having "something to show"), inertia in the sense of previous solutions and existing vested interests, and, above all, schedules.

Evaluation: Validity vs. Acceptability

Evaluation is the process of demonstrating that a problem has been solved. Researchers tend to view evaluation as an *opportunity to assess the validity* of the proposed solution, usually relative to alternative solutions. In contrast, designers typically look at evaluation as a *requirement to assure the acceptability* of the proposed solution. Succinctly, researchers are concerned with proving that a solution is "the best," whereas designers have to assure that a solution is "good enough."

While this difference between research and design may seem undesirable, it is actually quite natural. By focusing on scaled-down, context-free problems, researchers' pursuit of validity is quite feasible. On the other hand, designers' constraint-driven emphasis on context and contingency-laden problems causes the primary concern to be with solution acceptance on the part of management and eventual users.

Implications

Based on the above analysis, it should not be surprising that many problems and impediments are encountered in transitioning technology from

R&D to design. The process is inefficient and often ineffective. The question, of course, is what can be done to remediate these deficiencies.

It appears that a slight reorientation of the research and design communities could lead to substantial improvements. First, the research community should be more application-oriented in the sense of justifying their research in terms of its *long-term* (e.g., 5, 10, or 20 years) implications and communicating their results so that the implications for applications are readily apparent. This change could easily be accomplished if research funding organizations made treatment of these issues prerequisites for funding.

Similarly, the design community should adopt a *much* longer planning horizon (e.g., 5 or 10 years) so that the potential value of research results becomes an element of their investment strategy. The evolutionary architectures concept discussed in Chapter 5 provides a framework within which product generations can be conceptualized and long-term R&D requirements identified. This change can only be accomplished if management decides that long-term payoffs have value and if it rewards people (researchers and designers) who contribute to assuring that long-term payoffs emerge. Consideration of the impact of an organization's planning horizon leads to discussion of the other end of the technology transition process.

TRANSITION FROM DESIGN TO THE MARKETPLACE

Much has been said and written about the global marketplace and the ability of the United States to compete in this marketplace. The United States is a leader in *invention,* but not in moving inventions throught the *innovation* cycle to beat the competition. Why not?

One might pursue this question by studying economic and political theories, by manipulating various models of supply and demand, exchange rates, technology infusion, etc. An alternative approach—in the spirit of the naturalist phase—is to talk to the people most directly involved with the problem.

To this end, a workshop was conducted involving over 30 participants, mostly from industry (82 percent). A substantial majority of participants were engineers and scientists (77 percent) from higher levels of corporate management (69 percent). A structured workshop process was used, involving four independent working groups that independently pursued a series of questions. These questions focused on defining the problem, determining solution alternatives, and considering implementation strategies. This section discusses the results of this workshop on technological innovation (Rouse and Rogers, 1990).

Problem Definition

There was virtually unanimous agreement that the world has changed and there is now a global economy. However, we do not really understand the dynamics of the new world. This includes relationships among variables, parameters within relationships, rules of the game, and priorities.

Further, the U.S. role has changed. Indicators of these changes include the negative trade balance, lack of productivity growth, declining non-defense R&D spending, and the prevalence of industrial "downsizing." This raises the question of what the role of the United States should be, and how this role relates to the nature of competitors' roles.

With these broad changes have come more specific changes in terms of new standards. Quality is now critical, as are unit production quantities and short product life cycles. While it is possible to meet these new standards (e.g., Japan and Germany), there is no obvious panacea.

Big business in the United States is having difficulty with these standards. In part, this may be because big business is very risk-averse, perhaps because of the relatively high cost of capital. However, the answer does not seem to really be that simple.

Solution Alternatives

Participants concluded that solutions should remediate four particular aspects of the problem:

- Inappropriate incentives,
- Inadequately educated workforce,
- Adversarial government–industry relations, and
- Preponderance of barriers.

Uncertainty about incentives was a recurring theme. How can innovation be made worthwhile? What incentives are necessary to increase interest in pursuing technically oriented education?

The relative roles of the players are also not clear. To what extent should government and academia contribute to the solution? How should the risks of innovation be apportioned between the financial community and industry?

There was a consensus on emphasizing incentives and relationships. Incentives should be improved and barriers eliminated both internally (within industry) and externally (from government). Supportive relationships should be developed among industry, government, and academia. More emphasis should be placed on the central value of teamwork.

It was also felt that the United States needs a greater technological orientation. The status of technology should be improved in both business and government. Business plans and technology plans should be coupled. The educational system should be improved in terms of both technology and business orientation.

A key element for all of the above to succeed is the will and ability to make long-range commitments. While there is great uncertainty due to the changing nature of the world, an important element of uncertainty is our unwillingness and/or inability to choose investment plans, tax policies, and so forth, and remain committed for a length of time sufficient to determine if they work or not.

Implementation Strategies

Regarding the implementation "track record" to date, several observations emerged. First, there has been a proliferation of specialized centers and R&D consortia. There are also an increasing number of data bases available to support gathering of business intelligence. But despite all this activity and information, there is an unfortunate bias against reporting and compiling information about failures—it is difficult to learn from failures if we cannot admit that they occur!

Moreover, we have surprisingly little understanding of the true costs and benefits of specialized centers and R&D consortia. We really do not know if such endeavors are a good investment. We also do not know how to sift through all the data available to gain the information needed.

There was a clear consensus that alternative implementation strategies have strong implications for management. There is a need to develop approaches and methods for management of technology in both industry and government. Such developments would, for instance, result in recognizing technology transition as a real task, rather than something that is assumed to happen on its own. Technology transition should be programmatically managed and not rely on serendipity.

There is also a need to develop information management technologies that help in gathering and sifting through business intelligence information. In addition, business and technology plans should be more integrated, as well as clearly and widely communicated. Finally, as noted earlier, it is also necessary to create appropriate incentives and eliminate barriers.

Implications

It is important to remember that the above observations and assertions are the *opinions* of more than 30 technically oriented, mid-to-high level executives and managers. While these opinions are informed and educated, they are nonetheless subjective in nature. From this perspective, these results are best viewed as the concerns of those people who are charged with dealing with the technology aspects of the competitiveness issue.

Based on these concerns, it is quite clear that the potential for human-centered design to provide successful products and systems depends to a great extent on management philosophies and policies. For example, pursuit of the naturalist and marketing phases requires an early investment, perhaps years before the product or system is in the marketplace. Without a long enough planning horizon, this investment may not seem worth it.

As an even more important illustration, the methods described in this book work best if at least a few members of the design team are involved throughout the whole life cycle of the product or system. Accordingly, people involved in the naturalist phase should also participate in the sales and service phase. This requires management commitment to enable designers to cross functional areas within the organization, as well as spend periods of time watching and listening rather than designing.

Finally, and of great importance, technology transition requires that management recognize and commit resources to enable designers to learn about, appreciate, and deal with the plethora of people-related issues that pervade human-centered design. Put simply, designing for success requires that designers understand the players in the game, the roles they fill, their disciplinary perspectives, and how they perceive risks and rewards. To make these ideas more concrete, the next section presents an analysis of experiences in transitioning technology to manufacturing applications.

INFLUENCE OF ROLES AND DISCIPLINES

This analysis is based on a three-year effort to transition advanced interface technology, such as discussed in Chapters 8 and 9, from the aerospace, marine, process, and power domains where they were originally developed to manufacturing applications (Rouse and Hunt, 1990). The results of this process have been mixed, with some false starts and a few modest successes. However, in the context of technology transition, these experiences have provided important insights into how people in different roles and disciplines influence the process.

The people from industry encountered in the pursuit of this effort varied in terms of both jobs (or roles) and education (or disciplines). Roles include:

- Technically oriented senior *executives* responsible for assembling and managing technology portfolios,
- Middle *managers* responsible for particular programs or larger projects, and
- Technical staff members *(engineers)* responsible for smaller projects or tasks within larger projects.

The people within these roles are usually trained within one of the following disciplines:

- *Management* with primary interests/expertise in investment, risk, and return,
- *Industrial Engineering* with primary interests/expertise in process planning, material flow, and productivity, and
- *Mechanical Engineering* with primary interests/expertise in product geometry, machine dynamics, and performance.

Figure 10.2 summarizes the perspectives of people in the above roles and disciplines within manufacturing. The contrasts presented in this figure provide an explanation for an experience that has happened repeatedly. Typically, initial contacts in manufacturing organizations have occurred with either senior executives looking for innovations, or engineers interested in advanced aiding and training technology. Thus, initial contacts tended to be via the top and/or bottom of the organization.

DISCIPLINES (EDUCATION)	ROLES (JOBS)		
	EXECUTIVE	**MANAGER**	**ENGINEER**
MANAGEMENT	• Focuses on business innovations • Hedges investments across business opportunities • Tends to initiate new business directions • Success not tied to particular projects	• Focuses on risk reduction • Concerned with return on investment • Tends to delay and/or veto new directions • Success tied to particular projects	• This combination seldom occurs
INDUSTRIAL ENGINEERING	• Focuses on process innovations • Hedges investments across technology opportunities • Tends to initiate novel technology programs • Success not tied to particular projects	• Focuses on risk reduction • Concerned with productivity impact • Tends to delay and/or veto novel technologies • Success tied to particular projects	• Focuses on process technology • Concerned with process, flow, etc. • Tends to advocate novel process technologies • Success tied to technical tasks
MECHANICAL ENGINEERING	• Focuses on product innovations • Hedges investments across technology opportunities • Tends to initiate novel technology programs • Success not tied to particular projects	• Focuses on risk reduction • Concerned with performance impact • Tends to delay and/or veto novel technologies • Success tied to particular projects	• Focuses on product technology • Concerned with geometry, dynamics, etc. • Tends to advocate novel product technologies • Success tied to technical tasks

Figure 10.2. Influence of roles and disciplines.

Eventually, however, a middle manager became the customer in the sense that he or she had to either take the "hand-off" from the executive, or approve what the engineer was advocating. This stage is critical because middle managers have a tendency to regard new technology as risky and not necessarily solving today's problems. They would much prefer off-the-shelf technology that has been proven via many previous applications. In addition, probably due to pressures from higher levels of management, middle managers tend to focus on short-term, quantifiable cost savings and/or improvements.

		TIME FRAME	
		PRESENT	**FUTURE**
ASSUMPTIONS ABOUT PRACTICES	**BUSINESS AS USUAL**	Today's practices with today's technology	Today's practices with tomorrow's technology
	BUSINESS AS POSSIBLE	New practices with today's technology	New practices with tomorrow's technology

Figure 10.3. Alternative business perspectives.

Business Perspectives

As a result of the differing points of view among roles and disciplines, there can be differing perspectives on business, as illustrated in Figure 10.3 (Rouse and Cody, 1989). The resulting mismatch of perspectives can be summarized as follows:

- Executives are concerned with business as it might be,
- Engineers are concerned with technology trends and opportunities,
- Advanced technology such as discussed in Chapters 8 and 9 focuses on the confluence of these two concerns, and
- Managers are concerned with business as it is and how off-the-shelf, proven technologies can make incremental improvements.

Based on Figures 10.2 and 10.3, it should be fairly clear why technology transition tends to be difficult.

Implications

The contrasts presented in this section illustrate a central value of human-centered design. By developing and maintaining an awareness of the perspectives of the many stakeholders in design, on both the producing and consuming sides, one is in a position to know what matters most to each stakeholder, as well as know how to explain and advocate design efforts in appropriate terms. Of course, the central value of human-centered design is that the resulting design will, to the extent reasonable, incorporate and integrate the needs and preferences of all major stakeholders. This greatly increases the chances of technology transitioning appropriately, and products and systems being successful in the marketplace.

PREREQUISITES FOR SUCCESS

How can the technology embodied in this book be transitioned to successful use? More specifically, what are the conditions under which human-centered design is most likely to be successful? As noted in several places in this chapter, management commitment and support are necessary conditions. However, there are four other prerequisites (Rouse, 1987).

Long-Term Perspective

A long-term perspective is needed to balance and integrate measurement issues. For example, planning for viability, acceptability, and validity measurements, for which closure will not be reached for several years, requires a long-term commitment. In contrast, the lack of a long-term perspective often results in measurement being a victim of resource constraints.

Sense of Accountability

Both ethical and legal accountability are needed throughout the design life cycle. This motivates designers, as well as managers, to assure that they are meeting the needs of users, customers, and other stakeholders. Without this accountability, individuals pursuing one phase of design tend to feel that issues associated with later phases of design are "not their problems."

Flexible Design Process

Since designers are rarely clairvoyant, it is necessary to have a flexible design process that enables feedback of measurements into design refinements prior to production. The approach described in this book, including measurement concepts, methods, and guidelines, provides the basis for the requisite flexibility. Without this flexibility, feedback may not be possible or at least not meaningful. As a result, the motivation for measurements may be limited to satisfying contractual requirements.

Cooperative User–Producer Relationships

These relationships are the key elements in providing the necessary flexibility. The naturalist and marketing phases afford opportunities for building relationships, and the sales and service phase supplies the means for maintaining them. As noted in earlier discussions in this chapter, cooperation based on such relationships can provide the foundation for design success. Unfortunately, the adversarial nature of user–producer relationships in many domains presents a strong disincentive for planned measurement.

PROSPECTS

The case studies discussed throughout this book should make it obvious that human-centered design can provide innovative products and systems. Further, the concepts, methods, and guidelines discussed are not especially abstract or difficult to employ. Therefore, the elements of success are available and have been demonstrated to work. The prerequisites for realizing the full benefits of this approach are, admittedly, not necessarily easy to fulfill. However, the potential rewards are very great—innovative, quality products and systems that meet people's needs and succeed in the marketplace. Such results will no doubt include impressive sales and profits, but at least as important to the users of this book will be the satisfaction of design success.

REFERENCES

Rouse, W. B. (1985). On better mousetraps and basic research: Getting the applied world to the laboratory door. *IEEE Transactions on Systems, Man, and Cybernetics, SMC-15*, 2–8.

Rouse, W. B. (1987). On meaningful menus for measurement: Disentangling evaluative issues in system design. *Information Processing and Management, 23*, 593–604.

Rouse, W. B., and Cody, W. J. (1989). Information systems for design support: An approach for establishing functional requirements. *Information and Decision Technologies, 15*, 281–289.

Rouse, W. B., and Hunt, R. M. (1990). Transitioning advanced interface technology from aerospace to manufacturing applications. *Industrial Ergonomics,* to appear.

Rouse, W. B., and Rogers, D. M. A. (1990). Technical Innovation: What's wrong, what's right, what's next? *Industrial Engineering, 22,* 43–50.

Author Index

Subject Index